The Empty and the Full
Li Ye and the Way of Mathematics
Geometrical Procedures by Section of Areas

The Empty and the Full
Li Ye and the Way of Mathematics
Geometrical Procedures by Section of Areas

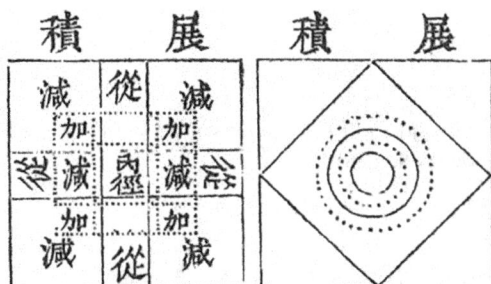

Charlotte-V Pollet

National Chiao-Tung University, Taiwan

W|≡ World Scientific

EW JERSEY • LONDON • SINGAPORE • BEIJING • SHANGHAI • HONG KONG • TAIPEI • CHENNAI • TOKYO

Published by

World Scientific Publishing Co. Pte. Ltd.

5 Toh Tuck Link, Singapore 596224

USA office: 27 Warren Street, Suite 401-402, Hackensack, NJ 07601

UK office: 57 Shelton Street, Covent Garden, London WC2H 9HE

Library of Congress Control Number: 2020011813

British Library Cataloguing-in-Publication Data
A catalogue record for this book is available from the British Library.

THE EMPTY AND THE FULL: LI YE AND THE WAY OF MATHEMATICS
Geometrical Procedures by Section of Areas

ISBN 978-981-120-947-5 (hardcover)
ISBN 978-981-120-948-2 (ebook for institutions)
ISBN 978-981-120-949-9 (ebook for individuals)

For any available supplementary material, please visit
https://www.worldscientific.com/worldscibooks/10.1142/11531#t=suppl

Acknowledgements

This book is the result of a project that started in 2007, when Karine Chemla suggested that I use my knowledge of Sanskrit and Chinese for history of mathematics. I was then teaching philosophy and was just a friend of mathematics, though I had studied Sanskrit and basic modern simplified Chinese at university. In order to improve my Chinese and mathematics, I was sent to the National Taiwan Normal University Department of Mathematics to study with Hrong Wann-Sheng. In order to teach myself mathematics in classical Chinese, I was also assigned Li Ye's *Yigu Yanduan*. We decided to build a four-year joint thesis partnership, which was the beginning of big adventure. The *Yigu Yanduan*, which was supposed to be a simple book and a simple chapter of my dissertation, had much more to offer than expected. It became the heart of my research. My dissertation contains an attempt of complete translation and some commented interpretations. I added new chapters and deep modifications to the present book.

I want to thank all those who accompanied me during these years and who made this possible: my advisors for their trust and patience. Any remaining mistakes are mine. I hope I did not betray them. I am indebted to Ken Saito for introducing me his computer program DRAFT, and to Guo Shuchun, Sun Chengsheng, Tian Miao, Han Qi for welcoming me to the Chinese Academy of Science in Beijing. It is my pleasure to express to these people my deepest gratitude. Thanks are due to Zhu Yiwen, Felix Zhen, Jeff Chen, Alexei Volkov, Andrea Breard, Agathe Keller, Christine Proust, Catherine Jami, Fu Dawei, Chu Pingyi, whose discussion enriched my work.

The writing of this book was interrupted by maternity and spine surgery. Wang Yi-Chih and Rachel Lee from National Chiao-Tung University played a key role in the elaboration of this book. They were literally my right hand during my recovery. I have pleasure here in expressing my deepest thanks to them as well as to my friend Ying Jiaming for his support. I express also my deepest gratitude to those who accepted the anonymous work of being referee.

The whole endeavor has been made possible thanks to the patronage of the Taiwan Ministry of Education and Taiwan Ministry of Science and Technology, whose financial help made everything possible. It is my pleasure to address my heartfelt thanks.

Conventions

Hanyu pinyin is used for transliteration of Chinese characters. The first occurrence of a Chinese term is accompanied by its traditional Chinese characters. Subsequent occurrences are rendered with their English translations. If the Chinese expression is older than the 1950s reform of simplified Chinese, then the expression is written in traditional characters. Simplified characters are used for modern names.

Different sentences within the same problem are numbered sequentially; for instance, in Problem 18, the different sentences are numbered [18.1] ⋯ [18.3] for the Chinese text, its translation and references.

Inside chapters, diagrams are numbered according to the number of the chapter. [Fig. 3.4] refers to diagram 4 of chapter 3. Diagrams provided in critical edition and translations at the end of each part are numbered according to problems. [Fig. P2.8] means diagram 8 of Problem 2.

Diagrams in black and white are reproduced with DRaFT program created by Ken Saito (Osaka Prefectural University). This program allows to reproduces diagrams exactly as they are in the Qing dynasty edition, with their imperfection. The coloured diagrams for the mathematical interpretation are creations produced with Geogebra.

Tabular arrays for polynomials are reproduced with a specific computer program, 'rod numeral generator' created by Wang I-Chih (National Chiao-Tung University, Taiwan).

Disambiguation: The expression Section of Pieces [of Areas] (no italic) denotes the geometric procedure. The expression *Development of Pieces [of Areas]* (italic) is the shortened translation of the title of the Chinese treatise containing the aforementioned procedure.

Preface

Jamais la connaissance n'est souveraine; elle devrait, pour être souveraine, avoir lieu dans l'instant[1].

Georges Bataille

The Song (960–1279 AD) and Yuan (1279–1368 AD) Dynasties China experienced a peak in high-level algebraic investigation, as exemplified by the works of famous mathematicians such as Qin Jiushao 秦九韶, Zhu Shijie 朱世杰, Yang Hui 楊輝 and Li Ye 李冶. Among these, Li Ye's short treatise *The Development of Pieces of Areas According to the Collection Augmenting the Ancient Knowledge* (*Yigu yanduan* 益古演段) elucidates a curious ancient geometrical procedure. Traditional scholarship has long discredited the importance of this book, castigating it as merely a popular handbook. Li Ye's work actually inaugurates a completely new aspect of ancient Chinese mathematics: a combination of algebra, geometry, and combinatorics, containing elements reminiscent of the *Book of Changes* (*Yi Jing*, 易經). Li Ye used field measurement as pretext for investigations on quadratic equations and changes but Li Ye's small treatise invited investigations into the link between mathematics and philosophy. The real but hidden topic is the exploration of expression of proof and generality in Chinese mathematics.

Li Ye (1192–1279 AD), a scholar and Song Dynasty official from northern China, is known for two mathematical treatises, the *Ceyuan Haijing* (測圓海鏡, 1248 AD) and the *Yigu Yanduan* (益古演段, 1259 AD). The first treatise is famous for being the earliest evidence of polynomial algebra in China. It has long been known thanks to its use of an algebraic method called the *Procedure of the Celestial Source* (*Tian yuan shu* 天

---

[1]BATAILLE, Georges. Ce que j'entends par souveraineté. Œuvres Complètes, vol. VIII, Gallimard, Paris, 1976, p. 253.

元術). *The Development of Pieces [of Areas]* presents the same method, but on a simpler level. For this reason, it has been considered a kind of introductory textbook, and thus inferior to the masterpiece. Yet, this traditional view of the *Development of Pieces [of Areas]* ignores the fact that the shorter treatise also contains another method, the geometrical procedure named Development of Pieces [of Areas] (*yan duan* 演段). Albeit mentioned rather briefly in the treatise, the presence of the *yan duan* is of immense importance.

Elements of the *yan duan* are found in another treatise from southern China, namely the second chapter of *Yang Hui's Methods of Computation* (*Yang Hui Suanfa* 楊輝算法1275 AD) by Yang Hui (second half of the thirteenth century). Both authors refer to more ancient sources. Li Ye refers to a source which can be traced back to the eleventh century (Jiang Zhou 蔣周 and his *Yiguji* 益古集) and Yang Hui cites a source from the tenth century (Liu Yi 劉益 and his *Yigu genyuan* 議古根源). A comparison of the procedure as described by Li Ye and Yang Hui helps to restore contemporary perspectives on of algebra and geometry in Song and Yuan Dynasty China.

The Development of Pieces [of Areas] draws on a long tradition of practices related to diagrams. The practices can be traced back to at least the Han Dynasty (206 BC–220 AD) in the classic *Nine Chapters on Mathematical Procedure* (*Jiu zhang suanshu* 九章算術, prior to the early first century AD) and its commentary by Liu Hui 劉徽 (third century AD). What is of particular interest here is the reliance on non-discursive practices to explain mathematics: drawings, visualising transformations of figures, and linking figures together according to criteria of resemblance. Li Ye seems to have borrowed this distinct methodology, which calls into question the author's intentions. How old is this structure? What is its meaning? Who is the founder of this methodology? Because of the absence of illustrated sources prior to this treatise, the *Development of Pieces [of Areas]* is the missing link which enables us to reconstruct ancient Chinese practices of geometry.

Far from a simple textbook, Li Ye's treatise is in fact an important and sophisticated treatise which investigates negative coefficients in second degree equations, solves combinatoric problems by means of geometry, and strives for generality as evidenced by the structure of the algorithms. This result is important because it is still commonly believed that "In China [···] there is no evidence of either geometric or algorithmic reasoning in the solutions of quadratic equations. All equations, of whatever degree above the first, were solved through approximation techniques" (Katz and Hunger

Parshall 2014, 8) or that "in mathematics nothing of value is achieved by changing the figures in a problem" (Libbrecht 1973, 4). The reader will see that the interpretation of ancient sources is subject to the understanding of mathematical practices, which can be far different from those to which modern readers are accustomed. An attention to mathematical practices is a necessary condition for a renewal of the history of mathematics and its philosophy.

My first intention was to publish a complete translation of *Development of Pieces [of Areas]*. The first attempt at a translation can be found in my Ph.D dissertation. However, upon completion, this project quickly appeared vain and useless. Who would read sixty-four problems which look more or less identical? The repetitive nature of the treatise is an obstacle to its legibility. As this research was aiming at translation, the result was consequently meant to be read. And this is precisely the problem. *Development of Pieces [of Areas]*'s greatest beauty is not about reading. It is about doing. The reader must first become a practitioner of this mathematical method, which requires him or her to make something in order to solve each of the sixty-four problems. The lesson of *Development of Pieces [of Areas]* is the construction of geometrical solutions to quadratic equations by means of imaginary transformations on diagrams. This mental practice implies inner transformations consisting of long-term, fundamental, cognitive changes.

Therefore, instead of producing a translation of Li Ye's treatise, I have chosen to exhibit directly the practices and their results. This book is about an experience to which the reader is invited. And this experience explores the limits of language which were met by Li Ye.

The Development of Pieces [of Areas] is composed of 64 problems with their solutions illustrated by one or two geometrical diagrams. A thematic reading may be based on a selection of problems. To this end, I have selected one problem to illustrate each chapter of the book. These problems were selected for their didactic and visual values. For each of the selected problems I have prepared a critical edition of the medieval Chinese text, a complete translation into English and a mathematical transcription. Each step of the solutions to these problems is illustrated by several diagrams created with the *Geogebra* computer program. These artificial diagrams are snapshots of moments in the transformations which the practitioner is supposed to imagine. The main thread of the present study, however, is an attempt to transform a mere list of problems into a narrative. This goal raises new questions: how may movement be communicated through fixed 2D diagrams? How can what the practitioner understands from this nonverbal practice be expressed verbally?

These concerns justify the 'impressionistic' style sometimes used in the representation of the text. It is difficult to verbalise such topics. After all, the goal is to communicate the impressions produced by visualisation as strong and immediate effects of perception on intellect. Li Ye transmitted a practice which produces a perception of generality through the visualisation of geometrical diagrams. These visualisations make the practitioner feel why a procedure is correct. The experience of repeating these shared visualisations leads to an unstated understanding of generality. As a series of impressions and perceptions, this communication extends beyond words. Therefore, the purpose of this book is to share an experience that was first entertained by a Song dynasty mathematician: proof without words.

To understand this experience, which consists of perceiving generality through the visualisation of geometrical objects, the reader must understand the order of the sixty-four problems. But to understand this order, the reader must first understand what geometrical object is to be transformed and how this object can be modified. To reach this step, we must first understand how diagrams are used. This book develops progressively from the diagrams to the order of problems. Thus, the present study is divided in three parts: diagrams, transformations, and order of diagrams. The three aspects of the treatise represent three didactic steps toward a shared experience.

If I discuss only these three aspects, it is because other topics have already clearly discussed elsewhere. Abundant literature outlines the procedure of Celestial Source, the use of fractions, decimals, rod numeral notation, etc. Revisiting these topics would be redundant. This book is not a detailed study of Song Dynasty mathematics, but an avenue to further study. This book represents an intersection of the history of mathematics, a description of meditative techniques as described by cultural historians, and the philosophy of language. From this vantage point, Li Ye exhibits a peculiar conception of generality and proof. It is too early to say if Li Ye was representative of a whole generation of mathematician but that is why this work is an avenue to future studies.

This book aims at unveiling an episode in the history of mathematics, but it also addresses a literary problem. How can that which is not written be explained? To explain what Li Ye wanted to transmit, I have borrowed from experiences better known from in literature than from the sciences. Where the scientific writing fell short, I had to turn to another style of writing. Thus, this book also draws on my past training as teacher of philosophy in France. The works of Merleau-Ponty, Derrida, Humboldt, Eco or the reading of Bataille, Genet and Breton have been useful to an understanding of Li Ye's experience.

Here, I use philosophy as a means and as goal. Some studies in the history of mathematics have contributed to our understanding of cognition as related to the visualisation of diagram and the ways of reading a text [Mancosu (2005)], and I attempt to use other philosophical reasonings to bridge contexts. The element that relates Li Ye's work to its social context is its underlying philosophy — more specifically, the Quanzhen Taoist current of philosophy as it has been described by Isabelle Robinet. Because there is a specific philosophy of language in this current of Taoism, it is possible to identify the way of reading expressions of generality and proof in Li Ye's work. Without this philosophical background, Li Ye's work seems to be reducible to practical recipes on basic algebra for the beginner. However, Li Ye's mathematical treatise is about an experience. The experience as a shared creation by Li Ye and his readers touches the limits of language and its philosophy. At the same time, this experience does not turn away from philosophy; on the contrary, this experience pushes philosophy to its end and demands a way of thinking that makes philosophy mute, as I have learned before from the French writers, and later from readings of the Quanzhen school. In consequence of this linguistic limit, nothing is written, but only drawn. The experience of knowledge takes place in an instant. This philosophy embedded in this mathematical context is a denial of the laborious verbal wisdom sometime named philosophy in the occidental world. Li Ye's mathematical treatise questions the limit of language and the virtue of nonverbal thinking.

[Chemla (2016), 9] already remarked that "actors do not always formulate their reflections on generality explicitly. [···] In some cases, background knowledge about actors and their immediate contexts, especially the scholarly culture in the context of which they operated, can complement the sources that come down from them and help us describe how they understood generality and how they worked with it". Moreover, Li Ye, like other philosophers from China, explains nothing. The reader must read between the lines. He or she has to read texts the same way as he or she looks at Chinese paintings: blank spaces are as significant as inked figures. Silence is as significant as words. Written script is the sign of something else which passes unwritten. The non-written part is necessary for the written part to exist. Reading Li Ye's mathematical text makes sense only after understanding this philosophy of language, where diagrams are between written discourse and non-verbal description.

The practice of visualisation was already presented as a key to reconstructing *fangcheng* practices in R. Hart's study of the *Nine Chapters*. He noticed that: "One of the most surprising results of the investigation presented here [in the Chinese Roots of Linear Algebra] is the importance of visualisation. The essential feature of linear algebra problem is [⋯] visualisation. For it is the visualisation of problems in two dimensions as an array of numbers on a counting board and the 'cross-multiplication' of entries there that led to general solutions of systems of linear equations not found in Greek or early European mathematics" [Hart (2011), 191]. The work of Li Ye employs the same process, namely visualisation.

Chemla ([Chemla (2005)], [Chemla (2006a)]) has already pointed out the importance of geometrical diagrams in the expression of generality in the *Nine Chapters* and the *Gnomon of the Zhou*. She shows how various algorithms emerge from the same transformation of on particular figure. However, between the *Nine Chapters* and Song Dynasty, there is a gap of more than ten centuries. Practices evolve. The practice of visualisation is still a quest of generality. The figure is fundamental, in the sense that the reasons of the various algorithm flow from its form and structure: it is the common principle and the incarnation of generality. Whereas in Han Dynasty a single figure was sufficient, Li Ye's treatise exhibits a sophisticated system which combines figures. The expression of generality is 'demultiplied'.

This perspective once again raises cognitive questions. What role do visual images and diagrams play in mathematical activity? In addition to cognitive issues one might also investigate how mathematical thinking might depend on the culture in which it is embedded. There is a pedagogical use of visualisation. In the *Development of Pieces [of Areas]*, diagrams cannot be replaced or re-ordered and the procedure lose its meaning without diagrams. They are necessary. The visual thinking cannot be replaced by non-diagrammatic thinking. It is irreplaceable in the sense that one could not think through the same proof by a thought process in which the visual thinking is replaced by some different kind of thinking. Here, visual thinking is the appropriate tool for generality.

One of the purposes of this book is to showcase on how an actor valued generality in relation to a specific context and how he worked with generality depending on the context in which he was operating. The way Li Ye explored generality helped to relate his work to a specific context.

Until now, the only context that was known was his life as a recluse. Now, it seems that his conception of generality and his practice of visualisation are directly linked with practices of Quanzhen Taoist context of this time. This shows that there is a diversity of expression of generality. There is no *a priori* meaning, which could be valid across contexts. There are only different ways of understanding the general principles in different contexts. The present study shows one of them.

Contents

Chapter 0

Introduction: General Presentation

自荀子《儒效篇》:「不聞不若聞之,聞之不若見之,見之不若知
之,知之不若行之;學至于行之而止矣。」

Literal translation: Not hearing is not as good as hearing, hearing is not as good as seeing, seeing is not as good as knowing,
knowing is not as good as acting; true learning continues up to
the point that action comes forth.

Another attempt at translation: I hear and I forget; I see and
I remember; I do and I understand.

Xunzi 荀子 (310 BC–237 BC),
the *Teachings of the Ru* (*Ru Xiao Pian* 儒效篇)

One of the best ways to understand the mathematical reasoning of
Chinese treatises is to work through translations, not because translations
transpose a discourse from one language to another, but because they fail to
do so. However, I am deeply indebted to many wonderful works translating
Chinese mathematical texts. Without them, this present study would have
not been possible. But the peculiarity of the Chinese treatise I present
here is that it reaches the limits of translation and of language. More
concretely, as a translator, I have tried to be faithful to what Humboldt
calls *das Fremde* in his preface to the translation of Aeschylus' *Agamem-
non*. There is transformation in translation. Meaning is always movement.
The strangeness of language can be seen strikingly in the work of translation here. Humbold uses the term *fremdheit* (translated as "foreignness",
or "strangeness") to describe the sense of reader has when the translator
seems to have chosen what sounds strange or odd, as though a mistake
has been made in translation. *Das Fremde* (translated as the "strange",
or "unfamiliar") describes the reader's sense of reading something that is
recognisable, that has been translated appropriately, but that gives the

feeling of reading that word for the first time [Eco (2004)]. Humboldt showed that a proper understanding of language in its original form is not transparent or unproblematic. *Das Fremde* always disturbs the reader. The present study aims to translate some written excerpts of a mathematical treatise, but also to translate the invisible part of the same treatise.

Translation is an impossible exercise. Some word and reasoning have no equivalent. Thus, to think about untranslatable terms is what makes them illuminating and interesting. Not only are there untranslatable words, but there are also objects that cannot be translated because they are not written. A text is not always a narration. Texts are evidence of cultural practices, and non-verbal knowledge. How, then, can a text be made to 'speak' about tacit knowledge? Here, this study builds on an impossible translation of a critical edition of a Chinese mathematical treatise which also includes mathematical commentaries. The Chinese treatise in question contains a collection of mathematical diagrams. How can diagrams be translated? How can a diagram — which by definition is a silent object — be made to speak about the practice which it represents, implies? Mathematics does not consist of only a transmission of lists of results students have to read. It is also made of shared practices [Netz (1999)]. Mathematics is something we actively do, with a shared set of actions. Thus, the term 'practitioner' is preferred over 'mathematician' or 'reader' here. None of the actors mentioned here officially had the status of mathematician and reading was a small part of their activity.

The Chinese treatise in question presents a list of repetitive problems on linear and quadratic equations. This book presents translated samples of the most interesting of these problems[1]. Each of the problems is translated twice: once literally and again into modern mathematical language. It is common to give a modern mathematical transcription in history of sciences; literal translations are less commonly in use. The exclusive use of modern mathematical terminology would unwittingly standardise the contents. Everything would be framed according to the criteria of contemporary mathematical reading. This uniformity would prevent an understanding of the specificities of the text. A mathematical object that is well known today may conform to several different practices or conceptions in the past. Additionally, mathematical objects are cultural products elaborated by the work of different cultures, which did not use the same concepts in equivalent ways. The reconstruction of the idiosyncrasies of

---

[1][Pollet (2012)] presents the complete translation.

some Chinese mathematics is the goal of this study.

In the modern transcription, all quadratic equations have the same form, namely $ax^2 + bx + c = 0$. An equation is a statement that involves data as well as an unknown variable. It is formulated to allow a practitioner to determine the value of the unknown. A quadratic equation is a statement containing only x and its square x^2, and no higher powers of the unknown. Treatises dealing with this object would therefore be reduced to sets of solutions and the establishment of quadratic equations that initially look the same... at first glance. Following [Chemla (1995)], consider what a modern mathematician would write as $ax^2 + bx + c = 0$. For contemporary mathematicians, this object contains multiple aspects; it can be considered to be an operation but it can also be thought of as an assertion of equality. In another respect, the relation represented by this equation can be approached in various ways so as to determine the value of the unknown quantity x. There are various kinds of solutions: those obtained by radicals, by numerical methods like the so-called 'Ruffini–Horner' procedure and by geometric techniques, among others. [Chemla (1995)] shows that this combination of diverse elements is not found in ancient documents, and *'they did not undergo linear development, whereby a first conception of equations would be progressively enriched until we attain the complexity of the situation sketched out above'*. On the contrary, Chemla finds various ancient mathematical writings wherein the elements distinguished above are scattered and dissociated, and other writings wherein some elements are combined. Therefore, it may be that the history of algebraic equations has to be conceived as a combination of several types of equations and alternatively as syntheses between some of these aspects, when they happen to meet.

Consequently, the concept of 'equation' may or may not refer to the same object as that in ancient traditions, the idea of 'unknown' may not relate to the same reality. To compare results that look the same is not sufficient. Nothing should supplant the analysis of a mathematical text as primarily a text. A text is not always a description or a presentation; it is evidence of results and concepts. It contains traces of activities linked to their interpretation. Thus, a focus on literal translation must precede the mathematical transcription, with the latter being necessary but not sufficient for a practical analysis. Literal translation takes into account the manifestation of mathematical objects inside the text and the relations invoked by the way of 'talking about' these objects. Some differences are perceptible in the ways of expressing, shaping, and structuring the discourse.

What can be recognised as the same object occurs with different statuses in various sources. The history of mathematics does not only describe the evolution of procedures to solve equations but also the meandering of the concept of equation itself.

0.1   Li Ye and the *Development of Pieces [of Areas]*

From the mid-first millennium BC onward, in China, a number of mathematical texts were used as textbooks in state educational institutions. These texts were concerned mainly with administrative affairs and calendar making. But during the Song (960–1279 AD) and Yuan (1279–1368 AD) dynasties, there existed mathematical texts not related to official education. For instance, Yang Hui 楊輝, Qin Jiushao 秦九韶, Li Ye 李冶, and Zhu Shijie 朱世杰 composed texts which did not address economical or astronomical matters and were never used for state examinations. These authors referred to even more ancient texts for which they purported to write their treatises in order to transmit and clarify their ancient mathematical procedures. The reader can find details in western language on the general history in [Martzloff (1987)], [Libbrecht (1973)] or [Li and Du (1987)]. Qin Jiushao's works records the solution of higher degree equations and the method of simultaneous congruence. Li Ye and Zhu Shijie treat problems on the elimination of one or more unknowns from a system of equations and higher degrees of simultaneous polynomial equations. Yang Hui's work principally reflects the situation of mathematics as it was commonly used. In the reign of the Northern Song Dynasty, the technique of wood-printing block was highly developed [Chia (2002)]. The treatise, object of the present study, should not be separated from the history of printing.

This book presents a case study of a Chinese text, the *Development of Pieces [of Areas] [according to] [The Collection] Augmenting the Ancient [Knowledge]* (*Yigu yanduan* 益古演段[2]; Later shortened in *Development of Pieces [of Areas]*), which was written by Li Ye 李冶 in 1259. It was published in 1282. Li Ye is one of the famous scholars of the Song-Yuan period[3]. His literary name was Renqing 仁卿, and his appellation was

---

[2]'The Development (演) of Pieces (段) [of areas] [according to] [the Collection] Augmenting (益) the Ancient (古) [Knowledge]'. The translation is justified later.

[3]The Song dynasty is divided into two distinct periods: Northern Song (960–1127) and Southern Song (1127–1279). Yuan dynasty: 1271–1368. Biographies of Li Ye can be found in English in [Mikami (1913), 80]; [Ho (1973), 313–320]; [Lam (1984), 237–239]; [Li and Du (1987), 114]; and in Chinese in [Mei (1966), 107]. His life has been the

Jingzhai 敬齋. He was born into a bureaucratic family in 1192. Originally, he was known as Li Zhi 李治. However, when he discovered that his name was the same as that of the Tang emperor whose dynastic title was Tang Gaozong 唐高宗, he changed it to Li Ye. In 1230, he passed the civil service examination and was offered a post in the government. However, his service was cut short when the district to which he was assigned fell to the Mongols in 1232[4]. He took refuge in the north and finally gave up hope of having an official career when the Mongols conquered the Jurchen Kingdom in 1234. In this impoverished situation, he devoted himself to studies. His first piece of work in mathematics was the *Sea Mirror of Circle Measurement* (*Ceyuan Haijing* 測圓海鏡) written in 1248. He continued to live as a scholarly recluse in the Fenglong mountains 封龍山 in Hebei 河北, where he probably received students for instruction. In this environment, he produced the *Development of Pieces [of Areas]* in 1259. By 1260, Kubilai Khan 忽必烈 had, on several occasions, approached Li Ye for advice on state affairs and astrological interpretations. When Kubilai ascended the throne, Li Ye was offered an official post, which he declined twice. He died in the mountains of Fenglong in 1279. Li Ye was also the author of the following works: *Supernumerary Talks* (*Fan Shuo* 泛說), *Commentary of Jingzhai on Things Old and New* (*Jingzhai Gu Jin Tu* 敬齋古今黈), *Collection of Works by Jingzhai* (*Jingzhai Wen Ji* 敬齋文集) and *Amendments of Books on the Wall Shelves* (*Bi Shu Cong Xue* 壁書叢削)[5]. Among them, only the *Commentary of Jingzhai on Things Old and New survives*, together with some quotations from the *Supernumerary Talks* contained therein. What happened to these books or why they disappeared is unknown [Ho (1973)]. According to the biography of Li Ye written in the *Official History of the Yuan* (*Yuan Shi* 元史), Li Ye asked his son to burn all of his works except the *Sea Mirror of Circle Measurement*, because he felt that it alone would be useful to future generations but the extent to which his wishes were fulfilled is unknown. Fortunately, the *Development of Pieces [of Areas] and the Commentary of Jingzhai on Things Old and New* survived the fire.

---

object of several notices since the Yuan Dynasty, with the first being written in 1370 in *Official History of the Yuan*, ch.160, and the last being written in 1799 in *The Inventory of Biographies of Scientists* (*Chouren Zhuan* 疇人傳) by Ruan Yuan 阮元. This material is not treated in this present work.

[4]During Li Ye's life, northern part of China was in the hands of the Tatar Jin dynasty (1115–1234), and the western part was occupied by the Tangust dynasty of the Xi xia (990–1227). Around 1230 both parts were conquered by the Mongols, who were on a constant menace to the Southern Song (1127–1279).

[5]Translation of titles by Lam Lay-Yong [Lam (1984), 239]

The willingness to preserve only his mathematical masterpiece can be interpreted as an attitude of Li Ye to favor of what we call 'mathematics'. But it could also be that Li Ye wanted to respect the Taoist doctrine that philosophy to be something that is neither to be said nor written. The very first sentence of the Taoist canon, Daode Jing 道德經, '*the way that can be said is not the eternal way, the name that can be said is not the eternal name*'[6], is often interpreted as a negation of all possibility that language might express philosophy. Therefore, it could be that the will of Li Ye was, paradoxically, also in favour of Taoism. There is another hypothesis. Li Ye's order to burn his works coincides with a quarrel between Taoists and Buddhists that started in 1225, and which ended with the defeat of Taoists. In 1257, before Li Ye decided to burn his books, the emperor Kubilai commanded that all Taoist books be burnt, with exception of *Daode jing*. [Volkov (2004)] suggests that scientific, mathematical, and astronomical knowledge was, at least partly, transmitted through Taoist networks and was not confined to institutions. [Horng (1999)] noticed several coincidences leading to the conclusion that Li Ye may have been a member of the Taoist sect of the School of Complete Perfection (*Quan zhen jiao* 全真教). In Li Ye's time, this sect spread from Shandong to Shanxi, Li Ye's area. After the invasion from the North, this sect offered shelter to Chinese nationals and recruited members largely issued from intellectuals and civil servants. The sect first received the protection of the emperor Kubilai before being criticised by Buddhists who complained of prosecution and abuse of power from official members of the sect. The texts produced by the sect concern the inner alchemy (*nei dan* 內丹). They synthesise several Taoist currents (breathing exercises, practice of visualisation and alchemy) and references to Confucian classics. They focus mainly on mental education and secondarily physical education. They make systematic use of trigrams and hexagrams [Robinet (1991)]. The school recommended ascetic reclusion and an impoverished life. Although only one of the philosophical books by Li Ye is still extant, the fact remains that the majority of his works were Taoist. The reading of the *Commentary of Jingzhai on Things Old and New* shows an obvious knowledge of Taoist philosophy and contains references to the inner alchemy. Surprisingly, the Taoist guidelines are of great help to reading the *Development of Pieces [of Areas]*.

There is no trace of the Song-Yuan edition of the *Development of Pieces [of Areas]*. The oldest available manuscript copy appeared in the eighteenth

---

[6]道可道非常道, 名可名非常名

century in the Imperial Encyclopaedia, the *Complete Library of Four Trea-suries* (*Siku quanshu* 四庫全書; later shortened in *Complete Library*). This Encyclopaedia was initiated by the Qianlong emperor (1736–1795) who or-dered local and provincial officials to search, catalogue, and copy all rare and valuable manuscripts held in libraries. It is a compendium of 3593 ti-tles. The copy found in the *Complete Library* is based on the one included in the *Great Canon of Yongle* (*Yongle dadian* 永樂大典) commissioned by the Emperor Yongle (1360–1424) and completed in 1408. The edition of 1408 is lost. An editor of the *Complete Library* added a first commentary, introduced by the character 'commentary' (*an* 案) in 1789. This commen-tary appears in two small columns inserted inside the text written by Li Ye and in a column at the end of the equation-solving procedure. The author of this commentary remains anonymous but could have been Dai Zhen 戴 震 ([Lam (1984), 240]) who, according to [Li (1955)], also commented on the *Sea Mirror of Circle Measurement*. Dai Zhen was commissioned in 1775 by the Qianlong emperor with the re-edition of the mathematics section of the *Great Canon of Yongle* in the *Complete Library*. Nonethe-less, the commentary added to the *Sea Mirror of Circle Measurement* does not support a comparison to the commentary found in the *Development of Section [of Pieces]*. Therefore, it is not possible to confirm Dai Zhen as the commentator. Nine years later, in 1798 the mathematician and philologist, Li Rui 李銳, added his own commentary to his printed critical edition of the mathematical part of the private collection, the *Collected Works of the Private Library of Knowing Our Own Insufficiencies* (*Zhibuzu zhai congshu* 知不足齋叢書). His edition is based on the *Complete Library* and other manuscripts [Pollet (2014)]. The latest edition of the Chinese text was produced by Li Rui in 1789 and is the one now used by historians. Its reprint was republished by Guo Shuchun in 1993 in *Source Materials of Ancient Chinese Science and Technology: Mathematics Section* (*Zhongguo kexue jishu dianji tong hui: Shuxue pian* 中國科學技術典籍通彙: 數學篇). More recently, a new edition of the *Development of Pieces [of Areas]* was published in 2009 by Li Peiye and Yuan Min [Li and Yuan (2009)]. This edition contains a short introduction, and a version of the text based on Li Rui's edition with several additions with punctuation and explanations in modern simplified Chinese and modern mathematical terminology for each problem.

The *Development of Pieces [of Areas]* presents a list of sixty-four prob-lems in three rolls. All of the problems seemingly relate to the same prac-tical topic: calculating the diameter or side of a field that contains a pond.

Each problem follows the same pattern and the treatise may be described as repetitive. However, the central topic of the *Development of Pieces [of Areas]* is actually the construction and formulation of quadratic equations derived from problems about squares, rectangles and circles. The peculiarity of this text is that it introduces and distinguishes two different methods for setting up quadratic equations: the first is called the Celestial Source (*tian yuan* 天元) and the second is named the procedure of Section of Pieces [of Areas] (*tiao duan* 條段). This second procedure is used to establish quadratic equations by means of geometry. A third procedure, which is also geometric, is added to twenty-three of the problems under the title of the Old Procedure (*jiu shu* 舊術). Li Ye states that these methods, or some of them, issue from older treatises. The purpose of his treatise is to disseminate them amongst all readers. Thus, the *Development of Pieces [of Areas]* can be considered primarily as a product of the standardisation of more ancient mathematical practices, rather than a revision of mathematics in Chinese. This makes the *Development of Pieces [of Areas]* seem highly representative of the variety of algebraic practices of the time and more deserving of attention. This treatise is interesting because of the variety of methods proposed to set up quadratic equations for each problem and also because it was considered to be accessible to any reader.

Indeed, each problem is composed the same way and gives the same type of solution. First, the statement of a problem, introduced by '*Let us suppose*' (*jin you* 今有) gives the area of a field less the area of a pond, as well as one or several distances, which are usually sides, diameters or diagonal measurements. The problems then require the reader to calculate other distances that were not given in the statement. It asks the question (*wen* 問) followed with the answers (*da* 答) given immediately afterwards. A diagram representing the field and the pond follows the statement, inside which one (or several) of the distances given in the statement or in the answer are drawn. Some of the dimensions are also written down as a caption.

The problem is solved according to the first procedure, the Celestial Source, starting with choosing the unknown and ending by establishing an equation that the unknown satisfies. The procedure describes how to find the coefficients of the different terms of the equation and gives a list of operations and manipulations on a counting support that lead to these coefficients. These coefficients are presented using tabular settings of two or three rows. A coefficient of the equation is set on each of the rows. The rows are ordered by degree, the top row being the constant term and the third one being quadratic term. The procedure ends with the statement of the equation. Li Ye does not describe how to solve the equation for x; rather, the reader is expected to know how to extract its root. There may be several possible roots to the equation, but Li Ye gives only one of them — the positive one. Li Ye may or may not have considered the other roots. The given root of the equation is one of the values of the unknown used to solve the problem. Li Ye ends this part of the problem by indicating how to find the other distances asked at the beginning of the problem through knowledge of the root of the equation.

Afterwards, Li Ye presents the solution by means of a second procedure: the solution by Section of Pieces [of Areas]. The general strategy of this second procedure is to derive the terms of the equation from geometry. This part contains first a description of each coefficient of the equation introduced by the sentence: 'Look for this (i.e. the unknown) according to the Section of Pieces [of Areas]' (*yi tiaoduan qiu zhi* 依條段求之). Li Ye indicates the operations by which the data presented in the problem are transformed into coefficients of the equation. Each coefficient is coupled with fixed positions on a counting support, namely, the 'dividend' (*shi* 實); the 'adjunct' (*cong* 從) and the 'constant divisor' (*chang fa* 常法). The translation of these terms results from a choice made by historians[7] who perceived a strong analogy between the procedure of division and the procedure of root extraction. But for the moment, the terms may be associated respectively, with the constant term, the linear term (x) and the quadratic term (x^2) of an equation.

Immediately after this first sentence, there follows a small portion of text composed of one, or sometimes two, diagrams and of an explanation titled *yi* 義, which has been translated here as 'meaning'. The 'meaning' has the shape of a small commentary for which the object is the diagram. It chiefly states how to identify the terms of the equation from the diagram. It is difficult to give a general description of this part, because each of the 'meanings' is specific to the case being treated. Yi changes from an algorithm to another, according to procedures inside of which operations are involved. It refers to concrete purpose of operations in the context of their use. This point will be elaborated later.

Twenty-three of the problems are presented along with a third method, which is called the Old Procedure. Only one of these twenty-three problems includes a diagram. Diagrams are mostly absent. Yet, as from the

---

[7][Chemla and Guo (2004)], [Lam and Ang (2004)], [Li and Du (1987)], among others.

vocabulary is almost the same as that used for the Section of Pieces [of Areas], it is possible to deduce that the procedure was also geometric. The Old Procedure is usually very briefly stated and has the same structure as the first sentence of the Sections of Pieces [of Areas]: only the operations constructing the coefficients are stated with reference to the same three positions of the counting support.

Some of the problems are also presented with variations of the procedures. These variations are introduced by the expression 'another method' (*you fa* 又法) in Problem 3; 40; 44 and 56. They are placed after the procedure for Section of Pieces [of Areas]. Problem 6 presents three different methods with a new diagram for each. Five other problems have peculiarities: Problem 11 is composed of two problems with two different statements. Problems 44, 59, and 60 are presented without any Section of Pieces [of Areas] procedure; however, they are not entirely deprived of a geometric procedure because they are accompanied by solutions using the Old Procedure.

0.2   State of the Art

The difficulty in reading the *Development of Pieces [of Areas]* is to clarify the relationship between the procedures. Why is there a juxtaposition of solutions for a same problem? It is to underline the equivalence of procedures or to show the diversity of approaches? Is it to show the geometrical origin of the algebraic procedure? Is it to expose the difference between an old and a new procedure? Is it to prove the correctness of a procedure? Several interpretations have been offered why Li Ye assembled these different procedures together, and a preponderance of them suggests the primacy of the Celestial Source procedure.

Lam Lay-Yong proposed two hypotheses for the composition of this text, either that '*the tian yuan was new and Li Ye took the opportunity to justify its algebraic reasoning by falling back upon the traditional equivalent geometrical meaning*' or that '*as an equation derived using the old method through the tiao duan concept was not easy to understand, Li Ye used the tian yuan method to elucidate the origin of the tiao duan method and explain it by means of clear geometrical figures*' [Lam (1984), 264]. That is to say, in the first case, the Celestial Source procedure was new and its efficacy needed to be demonstrated using a well-known ancient procedure by the Section of Pieces [of Areas]. In the second case, the Old Procedure and its derivative the Section of Pieces [of Areas] form had become obscure or may have been forgotten and needed to be re-explained by means of the familiar Celestial Source procedure. According to both hypotheses, the Section of Pieces [of Areas] procedure is assumed to be older than the Celestial Source procedure. In the first case, though, the Section of Pieces [of Areas] was well known, whereas in the second hypothesis, the Celestial Source procedure was the better-known one. The second hypothesis was espoused by Mei Rongzhao [Mei (1966), 143], but Lam Lay-Yong did not favour one over the other. Another study by Kong Guoping [Kong (1999), 173, 197] also suggested that the Celestial Source was derived from the ancient geometric procedure. All historians agree on the anteriority of the geometric procedure and on the focus on the Celestial Source.

In fact, the first to propose this interpretation was the Qing dynasty mathematician Li Rui himself. In his afterword to *Development of Pieces [of Areas]*, he claimed admiration for the Celestial Source procedure. He also explained that, due to the difficulties of the ancient procedure, Li Ye wrote '*the two books, Sea Mirror of Circle Measurement and Development of Pieces [of Areas], to give the groundings of the procedure [named] to set up the Celestial Source.*'[8] He even justified the two characters '*yan duan*' of the title: '*[the character] 'development' (yan) stands for 'development of the set-up of the Celestial Source' and 'piece' (duan) is for 'look for this according to the Section of Pieces [of Areas]':*'[9] According to him, the *yan duan* demonstrates that the two procedures are equivalent and the purpose is to introduce the Celestial Source procedure for perplexed readers.

Lam Lay-Yong and Mei Rongzhao agree that there is a difference between the Old Procedure and the Section of Pieces [of Areas] procedure: in their view, the part of the text containing the Section of Pieces [of Areas] was created by Li Ye himself, whereas the Old Procedure was copied from an older source. And thus, the twenty-three solutions named the Old Procedure are borrowed from an ancient source, as is implied by the name. Kong Guoping's argument is original in that he not only claims that the Celestial Source was new and was the method chosen by Li Ye to clarify an ancient procedure but also that the whole geometrical procedure was a method borrowed from a predecessor. The geometrical procedure is in fact presented with two different names, the Section of Pieces of [Areas] and the Old Procedure. That is, the Section of Pieces [of Areas] procedure and the Old Procedure are not so different from one another. This hypothesis differs from the one presented by Mei Rongzhao. According to Mei Rongzhao,

---

[8]'[…] 故所著海鏡, 演段, 二書, 竝以立天元術為根本.'
[9]'所謂演者, 演立天元, 段者, 以條段求之也.'

only the Old Procedure is borrowed from a predecessor. The present study supports the hypothesis of Kong Guoping and tries to identify other items which can be attributed to more ancient sources.

In the twentieth century, the various translations of the four characters of the title, *yigu yanduan* 益古演段, also reveal the multiplicity of interpretations of the status of the book. Was it a textbook, a theoretical treatise, or a text on applied mathematics? The oldest occurrence of the translation, '*Exercises and applications improving the ancient methods*', was proposed by George Sarton [Sarton (1927), 627]. This first translation shows that the *Development of Pieces [of Areas]* was considered to be a type of miscellany with practical objectives (*e.g.* surveying). Translations agree that the purpose was to improve an ancient method and to distinguish the 'old' from the 'new'. For example, the title was later translated as '*New Step in Computation*' by Libbrecht [Libbrecht (1973), 19] and by J.N. Crossley in [Li and Du (1987), 114]. Lam Lay-Yong [Lam (1984), 237] also proposed her own translation: '*Old Mathematics in Expanded Sections*' and more recently Liu and Dauben [Liu and Dauben (1993), 302] have proposed '*Illustrations and Computations of Ancient Mathematics*'.

Two recent translations into French offer a different interpretation. They take into account that Li Ye, in his preface and title, refers to an older book entitled the *Collection Augmenting the Ancient [knowledge]* (*Yiguji* 益古集), and that the expression '*yan duan*' names a type of procedure which serves as the topic of the treatise. Thus, the two French renderings are '*Le yan duan (dévelopment des pièces d'aire) du Yiguji*' by [Horiuchi (2000)][10] and '*Le déploiement des pièces d'aires pour la [collection] augmentant les [connaissances] anciennes*' by [Chemla (2001)].[11] However, these two French publications are not dedicated to the *Development of Pieces [of Areas]*; the former explains the procedure of Section of Pieces [of Areas] by means of a portion of *Yang Hui's Methods of Computation* (*Yang Hui Suanfa* 楊輝算法)[12] written by Yang Hui 楊輝 in 1275. The latter analyzes the change in the use and meaning of the character *tu*, 'diagram' from the Han dynasty through the Song Dynasty. These two French renderings correspond to an interpretation given by Li Di. In 1997, Li Di [Li (1997),

---

[10][Horiuchi (2000), 238]: 'The *yan duan* (development of pieces of areas) of the *Collection Augmenting the Ancient [knowledge]*' (orginal literal translation).
[11][Chemla (2001), 12–13]: 'the deployment of pieces of areas for the [collection] augmenting the ancient [knowledge]' (my literal translation). Her translation changed in 2016. [Chemla (2016), 15] reads: 'Deploying the pieces for the [collection] augmenting the ancient [methods].'
[12]Translation of titles by Crossley in [Li and Du (1987)].

237–238] , quoting [Lan and Hong (1985), 11–12], suggested that research on the procedure of the Section of Pieces [of Areas] might have been Li Ye's original objective in writing *Development of Pieces [of Areas]*; however, due to the variation in the approach, the procedure was not easy for Li Ye's contemporary readers to understand. Therefore, Li Ye decided to add more detailed solutions based on the Celestial Source procedure. Li Di noticed that as a consequence, the Section of Pieces [of Areas] procedure became secondary and even dispensable. Hence, he suggested further clarification of the Section of Pieces [of Areas].

Despite Li Di's interpretation and probing of the meaning of the title, the *Development of Pieces [of Areas]* has been considered '*a revision of a work for beginners in the 'celestial element method'* ([Li and Du (1987), 114]), a book '*devoted to the method of tian yuan shu[13]*', or '*an 'introduction' to the Sea Mirror of Circle Measurement*' [Dauben (2007), 327]. It is generally thought that Li Ye '*took the opportunity to explain the tian yuan shu method in a less complicated manner after finding his first book (the Sea Mirror of Circle Measurement) too difficult for people to understand*' [Ho (1973), 319]. While the focus remains on the procedure of the Celestial Source, the procedure of the Section of Pieces [of Areas] is neglected, sometimes even published without diagrams ([Dauben (2007), 329]; [Guo (2010), 370–373]) or alternatively not mentioned at all[Ho (1973), 313–320]; [Li and Du (1987)]). There are several reasons for this absence.

First, the Celestial Source procedure was described at a higher level of difficulty and sophistication in the other major mathematical work of Li Ye, *Sea Mirror of Circle Measurement*. The *Sea Mirror of Circle Measurement* is said to be the distillation of Li Ye's studies on the Celestial Source, whereas the *Development of Pieces [of Areas]* represents his effort to popularise the method ([Mei (1966), 147]; [Lam (1984), 247]; [Kong (1999), 173]). The Celestial Source procedure reported in the *Development of Pieces [of Areas]* is too simple to attract much attention. Therefore, the whole treatise has been considered simplistic.

Second, interpretations have focused on the Celestial Source procedure because of the preface written by Li Ye. Here, he blamed mathematicians for their unwillingness to share their knowledge. He reproached them for writing in such an abstruse manner that mathematical knowledge was not revealed[14]. From this, scholars have inferred that Li Ye wrote the *Devel-*

[13]Guo Shuchun's introduction to Ch'en Tsai Hsin (Chen Zaixin 陳在新) translation of Zhu Shijie's *Jade Mirror of the Four Sources* (Zhu Shijie 1303. I. 46). See [Hoe (2008)].
[14]今之為算者，未必有劉、李之工，而褊心踞見，不肎曉然示人，惟務隱互錯糅，故為溟涬

opment of Pieces [of Areas] to correct the prevailing trend and to show how easily mathematical techniques such as the Celestial Source procedure could be learned and mastered even by beginners. This interpretation of the preface is supported by the fact that a large proportion of the treatise repeats slight variations of the Celestial Source procedure, and that the remaining part contains mostly diagrams and small discourses on them.

The two characters *tian yuan* 天元 however, never appear in the preface. On the contrary, only the characters *tiao duan* 條段 are mentioned. Li Ye wrote that he modified another treatise by adding diagrams:

'*A book entitled Collection Augmenting the Ancient [Knowledge] was compiled recently with reshaped [solutions to geometric problems of] rectangles and circles. It is indeed equivalent to work by Liu Hui and Li Chunfeng. However, I detest its reserved style, and hence add detailed diagrams of how to reshape the Sections of Pieces [of Areas]. Isn't it a great joy that the book will thus be accessible with basic knowledge from ten to hundred?*'[15]

Indeed, the vocabulary denotes clarification, but none of the elements explicitly correspond to the *Sea Mirror of Circle Measurement* or an *a priori* intention to render the procedure of the Celestial Source accessible

---

黯黮，惟恐學者得窺其彷彿也。不然，則又以淺近猶俗，無足觀者，致使軒轅隸首之術，三五錯綜之妙，盡墮於市井沾沾之見，及夫荒邨下里，蚩蚩之民，殊可憫悼。'*On the other hand, contemporary mathematicians, who do not necessarily study as comprehensively as Liu Hui or Li Chunfeng, are narrow-minded and short-sighted. Instead of making it clear, they prefer rendering it as implicitly and intricately as possible in order to make mathematics appear opaque and obscure. They prevent even a glimpse of its simulation being caught by others. Otherwise, some of them opt to deal with merely the basic and well-known part that is not worth looking into. Consequently, the methods of the ancients Xuan Yuan and Li Shou along with combination and alternation of numbers by three and five* (三五錯綜) *become something with which everyone in the town can be self-satisfied. It is such a pity that they actually know just as much as ignorant villagers.*'

[15]近世有某者，以方圓移補成編，號「益古集」，真可與劉李相頡頏。余猶恨其悶匿而不盡發，遂再為移補條段細繙圖式. 使粗知十百者，便得入室啗其文，顧不快哉? Original punctuation and translation. The expression 知十百 may refer to the *Analects of Confucius*. V.9 reads '何敢望回. 聞一以知十. 賜也. 聞一以知二'. [Volkov (2006), 66] translates this as '*Comment oserais-[je me] comparer á [Yan] Hui ?! Hui [lui], entend d'un et à partir de cela [il] connait dix; [moi], Ci, [j'] entends d'un et à partir de cela [ne] connais [que] deux*.' Volkov suggested a mathematical interpretation of the sentence. The number 'one' was represented by a vertical rod on the computation board and the number 'ten' by a horizontal rod. Yan Hui, favourite disciple of Confucius, was able to increase from one power of 10 to another by changing the position of the rods, while Ci (another disciple) was only able to add one unit — the number 'two' being represented by two vertical rods next to each other. The same expression *zhi shi*, 知十, is found in Li Ye's preface, but here, the reader is presumed to understand the number 'ten' and how to increase its power.

to a large audience. In fact, whether the simplification of the Celestial Source procedure was the main topic of the *Development of Pieces [of Areas]* remains unknown.

The *Development of Pieces [of Areas]* has been interpreted by historians as an introduction to the *Sea Mirror of Circle Measurement*. By this interpretation, the text served as a mere popularisation and thus remained in the shadow of the *Sea Mirror of Circle Measurement*. The book is still considered to be a list of simplified examples that show the Celestial Source procedure. Currently, the *Sea Mirror of Circle Measurement* by Li Ye stands as the oldest transmitted Chinese text that uses polynomial algebra in order to solve problems. It is evidence of the intense mathematical activity that seems to have developed in northern China during the Song and Yuan Dynasties. Numerous studies have analysed this book and it is mentioned in every publication dealing with the history of mathematics in China.[16] The *Sea Mirror of Circle Measurement* is famous for its highly technical presentation of the procedure of the Celestial Source, used to set up polynomial equations in one to four unknowns. Ever since its re-edition by the Qing Dynasty editors, this treatise has been used to introduce this procedure. For instance Li Shanlan 李善蘭 (1810–1882) considered Li Ye's *Sea Mirror of Circle Measurement* to be the most influential Chinese text he had studied before learning 'Western' mathematics. Later, when he taught at the Beijing school of Foreign Language (*Tongwen guan* 同文館), Li Shanlan used the *Sea Mirror of Circle Measurement* to prepare students to master modern algebra in order to show the efficiency of the method and to link the new algebra with Chinese tradition [Tian (1999), 114–115], [Horng (1993b)]. This could thus lead the reader to think that the topic of Li Ye's work was limited to the Celestial Source procedure. Consequently, the conventional wisdom has held that Li Ye's purpose was to make a sort of presentation of this procedure. Chemla has recently offered a new reading of this mathematical treatise ([Chemla (2005)] and [Chemla (1993)]). Chemla showed that the content of this book challenges definition. If read as separate formulas and problems, their algebraic content is obvious, but Chemla argued that the grouping of formulas is meaningful and relies on mathematical knowledge not explicitly stated in the book. She

---

[16][Chemla (1982)], [Chemla (1993)]) and [Kong (1996)] focused exclusively on the *Sea Mirror of Circle Measurement*. [Mei (1966)], [Kong (1988)], [Kong (1999)] and [Guo (2010)], among others, devoted one chapter to the reading of *Sea Mirror of Circle Measurement* and a few pages to the *Development of Pieces [of Areas]* ([Mei (1966), 147]; [Lam (1984), 247]; [Kong (1999), 173]).

reconstructed this system and argues that groups of formulas reveal Li Ye's mathematical preoccupations, which had not been considered previously. She demonstrated that the content of the *Sea Mirror of Circle Measurement* exceeds the mere sum of all formulas and problems; in her analysis, the structure of the book reveals specific concerns of Li Ye. Chemla thus questioned the definition of the content of ancient mathematical texts.

Although the Celestial Source procedure appears in other Chinese mathematical works, Li Ye provides the earliest reference to its use. The absence of other sources has led some modern scholars to believe that Li Ye's works are '*the first truly algebraic works in China*' [Dauben (2007), 324]. The identification of Li Ye's Celestial Source procedure as the foundation of polynomial algebra in the Song dynasty seems to be an undisputed fact. It is the quintessential '*Chinese algebraic process of logically setting up algebraic expressions and finding a relation between these expressions to derive an equation*' [Lam (1984), 243]. On the basis of the existence of this procedure, historians describe the thirteenth century as the acme of algebra in China and exemplify algebra in China by this procedure alone [Li and Du (1987), Ch. 5].[17] This procedure is essential to understand the form taken by algebra in China, yet Chemla showed that it is a tool that might not have been the principle purpose of the treatise. The purpose of the treatise might be something else, just as might have been the case with the other treatise by Li Ye, the *Development of Pieces [of Areas]*. Although this work is concerned with the Celestial Source procedure, it also supplies evidence of a different mathematical procedure. Surprisingly, this other procedure, i.e. the Section of Pieces [of Areas], has disappeared from works on the general history [Pollet (2014)].

0.3 The Sources

A similarly named procedure appears in a treatise written by another mathematician, Yang Hui (second half of the thirteenth century). The second chapter of *Yang Hui's Methods of Computation* (1275), titled *Fast Methods of Multiplication and Division Related to [Various] Categories of Fields and [their] Measures* (*Tianmu bilei chengchu jiefa* 田畝比類乘除捷法, later shortened in '*Fast Methods of Multiplication and Division*')[18] presents a

---

[17]'*Finally, for certain thirteenth-century Chinese mathematicians, Li Zhi, Zhu Shijie and their emulators, algebra was the tian yuan technique*' [Martzloff (1987), 258].

[18]Translation of the title by [Volkov (2007), 445]. *Practical Rules of Arithmetic for Surveying* is the translation of the title by [Lam (1977)].

procedure called *yan duan* 演段. Although it is a similar procedure, the two treatises emphasise different ways of elaboration, as if there were several stages of elaboration. The stages illustrated in the *Yang Hui suanfa* seem a precursor to those in the *Yigu yanduan*. But the *Yigu yanduan* also contains another version of the procedure of the Section of Pieces [of Areas], named the Old Procedure. This version of the procedure seems to predate the one in the *Yang Hui suanfa*. This raises the question as to whether these versions are pictures of different stages of elaboration. Or are they several layers of composition that reflect a specific moment in a work dealing with algorithms and geometrical representations?

Both mathematicians, Li Ye and Yang Hui, refer to earlier sources for the Section of Pieces [of Areas] procedure. Li Ye presents his motivation for writing the treatise in his preface: inspired by an old book titled *Collection Augmenting the Ancient [Knowledge]*[19] he compared its contents to the works of the two famous commentators, Liu Hui 劉徽 (third century) and Li Chunfeng 李淳風 (seventh century), who clarified the Han dynasty classic, the *Nine Chapters on Mathematical Procedures* (*Jiu zhang suanshu* 九章算術).[20] Li Ye found the *Collection Augmenting the Ancient [Knowledge]* obscure and incomplete and therefore decided to revise it and add diagrams. The revised text, though, was not distinguished from the original. It is generally accepted that the 23 solutions titled Old Procedure originated in the *Collection Augmenting the Ancient [knowledge]*.[21] The preface written by the editor of the *Development of Pieces [of Areas]* in 1282 states that the *Collection Augmenting the Ancient [Knowledge]* contained 70 problems, but the preface of the *Development of Pieces [of Areas]* in the eighteenth century re-edition in the imperial encyclopaedia, the *Complete Library*, argues that the number was smaller, probably 64 problems. Because no edition of the *Collection Augmenting the Ancient [Knowledge]* is extant, the relationship between the 23 problems solved by the Old Procedure and the 64 total problems has never been elucidated.

---

[19]The title is translated as '*Continuation of the Ancients*' by [Hoe (2008), v] and '*collection of old mathematics*' by [Lam (1984), 239].

[20]I follow Chemla's translation of the title [Chemla and Guo (2004)], but will later refer to the classic as 'the Nine Chapters'. The book was compiled in the beginning of the Common Era, was considered a 'canon' of classical Chinese mathematics, and its format, terminologies, and categorisation of problems had been considered standard since the 7th century. It was also the object of several commentaries. Among them, two were transmitted along with the canon, one by Liu Hui (3rd C.) and the other by Li Chunfeng (7th C.). It is the most influential mathematical text in pre-modern China.

[21][Lam (1984), 241]; [Mei (1966), 140]; [Kong (1999), 174]; [Ho (1973), 319].

Neither the names of the author nor the date of compilation are known with certainty. According to [Mei (1966), 139], the book can probably be attributed to Jiang Zhou 蔣周, from Ping Yang 平陽, in the province of Shanxi 山西. Mei Rongzhao relies on two references to a book, the title of which begins with the two characters *Yigu* 益古. In a preface to *The Precious Mirror of Four Sources* (*Si yuan yu jian* 四元玉鑑) by Zhu Shijie 朱世杰, Zu Yi 祖頤 (1303) lists the works to which later readers are indebted for the knowledge of the Celestial Source procedure. The first book on the list, *Yigu* by Jiang Zhou, is mentioned as having elaborated the procedure.[22] At the end of the sixteenth century, the Ming mathematician Cheng Dawei 程大位, compiled a list of mathematical texts produced between 1078 and 1224. Among them, a book titled *Computing Method Augmenting the Ancient [Knowledge]* (*Yigu Suanfa* 益古算法), may have been the *Collection Augmenting the Ancient [Knowledge]*. Xu Yibao found another citation of the same person from the Song dynasty [Xu (1990), 67]. Chapter Fourteen of the *Explanations for the Titles in the Zhizhai Book List* (*Zhizhai shu lu jieti* 直齋書錄解題, 1244) by Chen Zhensun 陳振孫 mentions the *Computing Method for Application* (*Yingyong suanfa* 應用算法), written before 1080 by Jiang Shunyuan 蔣舜元 from Ping Yang. Xu Yibao argues that Jiang Zhou and Jiang Shunyuan are two names of the same person.

Yang Hui presents extracts of an older, lost work titled *Discussion on the Origin of Ancient Methods* (*Yigu genyuan* 議古根源)[23] written by Liu Yi 劉益. According to Yang Hui, the *Discussion on the Origin of Ancient Methods* contained one hundred problems.[24] Yang Hui repeated 21 of those problems, 14 of which contained solutions called 'development of pieces', *yan duan* 演段. Liu Yi lived in Zhongshan 中山, in the province of Hebei 河北, during the late tenth or early eleventh centuries [Te (1990), 56]. On the strength of a comparison between the problems of Yang Hui's *Methods of Computation* and those of the *Development of Pieces [of Areas]*, [Xu (1990), 72] concluded that the *Collection Augmenting the Ancient [knowledge]* was probably composed in the middle of the eleventh century and was based on the *Discussion on the Origin of Ancient Methods*. These observations extend the procedure of the Section of Pieces [of Areas] to the eleventh century.

---

[22]Preface to *Zhu Shijie* by Zu Yi, translated into English by Ch'en Tsai Hsin (Chen Zaixin, 陳在新) 1925 [Zhu (1303)]. Reedited and completed by (Zhu Shijie 1303).

[23][Horiuchi (2000), 238]. Title translated into French: '*Reflexion sur les fondements des méthodes anciennes*'.

[24]'議古根源故立演段百問'.

The procedure of the Celestial Source, also used by Li Ye, is not used in the extant books of Yang Hui. Among the authors of mathematical texts written in Chinese, only Yang Hui and Li Ye recorded diagrams and substantial evidence of the Section of Pieces [of Areas] procedure.[25] It is difficult to determine the origins of this procedure or how widely it was used in the Song Dynasty. Li Ye, Jiang Zhou and Liu Yi were all natives of northern areas overlapping with the modern provinces of Hebei 河北 and Shanxi 山西, but Yang Hui hailed from Zhejiang 浙江 in southern China. Liu Yi and Jiang Zhou lived near Bianjing 汴京, the capital of the Northern Song Dynasty (960–1127). After the Song Dynasty lost control of northern China to the Jin Dynasty, the capital moved to Linan 臨安, now Hangzhou in Zhejiang Province in 1127. Li Ye remained in the north after the conquests of the Jin and Yuan Dynasties and lived as a recluse in the Fenglong Mountains. The political conflicts of that time strongly suggest that Yang Hui and Li Ye never met. Apart from Volkov's study on Taoist connections [Volkov (2004)], no information elucidates a network of mathematicians of that time or their libraries. If the works of Liu Yi formed part of the imperial library, its transfer from the old capital to the new one could explain how Liu Yi's mathematics reached Yang Hui. However, several studies [Zürcher (2007)] show that distance is less an obstacle to the probability of transmission than natural obstacles or the facility of trade routes. Bianjing is separated from Linan by only 875 km of central plains, across the Huai and Yellow river, and one or two minor mountains. [Morgan (2015)] shows the existence of a connected elaborated network of personal, professional, educational, and family ties extending across institutional, geographic and war-torn political divides in early imperial China, by example of the mathematical astronomy from the Cao-Wei 曹魏 court (220–265). This shows that as texts circulated, they evolved as well as did the knowledge they propagated. There was probably such a network in the Song dynasty as well, but its reconstruction is still pending.

There is almost no information on the context of the writing of either the *Development of Pieces [of Areas]* or Yang Hui's *Methods of Computation*. No information on Yang Hui survives and Li Ye is known only to have lived as a recluse when he wrote his treatises. He probably had disciples. However, nothing concerning Yang Hui is known. Whether these books were used for teaching and the intended readers are equally unknown. Concern-

---

[25]The term *yanduan* occurs three times in the opening pages of the book by Zhu Shijie (1303), when the author compiles the various technical terms derived from a right-angled triangle. [Bréard (2000), 266].

ing the *Development of Pieces [of Areas]*, none of the elements confirm the
link with the *Sea Mirror of Circle Measurement* and a popularisation of
the Celestial Source procedure. The simplification of the Celestial Source
procedure may not even have been the main topic of the *Development of
Pieces [of Areas]*. The purpose of the present study is to question this
point of view and to cast light upon a peculiar field in the history of Chinese mathematics. The *Development of Pieces [of Areas]* is a treatise on a
mathematical object which was new during the Song–Yuan period and for
which mathematical practices can be related to the famous Han Dynasty
classic, the *Nine Chapters*. The focus must be redirected to the other procedure: the Section of Pieces [of Areas]. This procedure concerns practices
of geometric diagrams and that these practices were not new by the time of
Li Ye. It is possible to recognise that more than just the 23 solutions using
the Old Procedure were borrowed from ancient sources. Surviving mathematical books prior to the Song dynasty are deprived of their geometrical
illustrations. In this context, the *Development of Pieces [of Areas]* becomes
a precious source; it suggests continuity and change within a tradition of
practice based on diagrams.

The *Development of Pieces [of Areas]* contains a collection of 131 diagrams. The concept of geometrically formulating an equation reached a
high degree of sophistication. In fact, the numerous diagrams constitute a
special and important feature of this book. However, all available editions
of mathematical treatises from the Song dynasty are products of Qing Dynasty critical works. It is therefore not possible to determine the extent
to which the available Qing editions are faithful to their sources [Pollet
(2014)]. Nonetheless, one of the available editions of Yang Hui's diagrams
preserves the earliest evidence of this procedure. It is a manuscript of a
Korean edition produced in 1433 based on a 1378 Ming edition, which was
found in Japan and brought back to Beijing at the beginning of twentieth
century and reprinted by Guo Shuchun in 1993. If the available editions of
other mathematical treatises establish the presence of diagrams included in
Song Dynasty treatises, they also show that treatises before the Song were
not provided with diagrams at all. Texts printed before the Song Dynasty
have not been passed down to us with their original diagrams, and the diagrams in subsequent editions were done later. Li Ye himself stated in his
preface that he added the diagrams. The *Collection Augmenting the Ancient [Knowledge]* probably did not contain published diagrams. The Song
Dynasty was a turning point in the policy of publishing diagrams. From
the available materials, some of their ancient features can be deduced from

the available texts. The diagrams related to the Section of Pieces [of Areas] procedure are part of a long tradition presenting the evolution of peculiar features.

A tradition of geometrical artefact has been identified dating back to the third century in a commentary by Zhao Shuang 趙爽, on the *Classic of Gnomon of the Zhou* (*Zhoubi suanjing* 周髀算徑)[26] and by Liu Hui in a commentary on the *Nine Chapters* [Chemla (2001)]. Neglecting the abundant literature which has been produced on the *Gnomon of the Zhou* is impossible. Since the earliest extant mathematical texts from ancient China, ranging from the second century B.C.E to the first century C.E, contain almost no information about the kind of visual aids practitioners of mathematics were using at the time that these writings were composed, the *Gnomon of the Zhou* holds a place of pride in Chinese mathematical literature. Its opening chapter contains a development some scholars identify as the first extant reference to graphical elements. The first commentary produced in the third century by Zhao Shuang presents a long development of the right-angle triangle and opens this development with three figures. The earliest pictorial evidences are found in a Song Dynasty edition (1213) [Fig. 0.1].

The goal is not to offer a thorough review of what has been published on this topic but rather to sketch a few lines on recent literature. Modern scholars still maintain divergent views on the dating of the *Gnomon of the Zhou*, the mode of composition of the text, and its epistemological status.[27] Cullen states that '*there is nothing in the main text that could be considered an attempt of proof*' [Cullen (1996), 87–88], while [Chemla (2005)] interprets the opening section of the *Gnomon of the Zhou* as '*relating to a statement and a proof of the 'Pythagorean Theorem''* [Chemla (2005), 127] relying on the Li Jimin restoration of diagram [Li (1982)]. [Qu (1997)] demonstrates that the commentator on the *Gnomon of the Zhou*, Zhao Shuang, not only recognised the stated proof but also reconstructed it in the form of a hypotenuse diagram. The commentator states his own proof of the Pythagorean Theorem. [Chemla (2005)] shows that there is a contrast between the diagrams referred to in the discourse on the *Gnomon of the Zhou* and the ones referred in its commentary. Chemla shows that there is not '*a unique figure on which to follow the text*' [Chemla (2005), 138], as is found in Euclidean geometry. The shapes were designed to support the argument evolved through the text, with figures being brought

---

[26]Later shortened to 'Gnomon of the Zhou'

[27][Qian (1963), 4]; [Cullen (1996), 139–156]; [Needham and Wang (1954), 19–20].

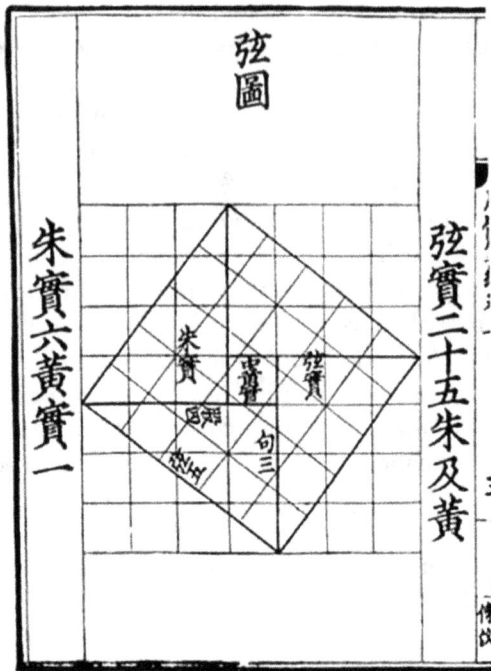

Fig. 0.1   *Gnomon of the Zhou* Diagram in the Edition of 1213.

together, cut, and moved. Chemla deduced from these features that these constitute a new type of figure that emerged in the third century and stated that the main difference between the first discourse and its commentary is that '*the figure is no longer an object that is created in relation to an algorithm and reshaped while proving its correctness. It is completed before the statement of any algorithm*' [Chemla (2005), 144]. The unchanged figure is then used to interpret the results of successive steps of procedures. It was a new basis for establishing in a uniform way the correctness of several algorithms. This change in the use of diagrams between the first and third centuries was due to emphasis on generality. She demonstrates that the same phenomenon can be observed in the famous mathematical classic, the *Nine Chapters* compiled in the first century B.C.E or C.E, whose commentary was produced in the third century by Liu Hui.

The classic itself does not contain reference to any specific geometric representation, but the commentary to the ninth of the *Nine Chapters*, which enunciates what is commonly called the Pythagorean Theorem under the title '*base (gou* 勹) and *height (gu* 股)', contains reference to visual artefact (to which it refers by the specific name *tu* 圖) and colors (Chapters 1, 4 and 9). This is the one of the principal topics of the *Gnomon of the Zhou*.[28] Colors are mentioned in the commentary to Chapters 1, 4 and 9 of the *Nine Chapters*, as they are in the commentary on the *Gnomon of the Zhou*. The two commentaries from the third century by Liu Hui and Zhao Shuang present some obvious relations. They reveal a circulation of knowledge [Chemla and Guo (2004), 662–710].

According to [Chemla (2010)], the third century *tu* were material objects, cut in paper with a square grid and worked in specific ways to be rearranged. They probably displayed particular dimensions and presented objects in plane geometry. The third century commentaries suggested that diagrams were already marked by characters or colored before the commentaries were composed. They were separate objects and this explains their absence from the texts. Another possibility is that collections of the diagrams were once published as separate booklets that were later censored. [Volkov (2007), 429–432] listed eight books that may have contained geometric diagrams. Several edicts from the Tang, Zhou and early Song Dynasties proscribed astronomical and divinatory books ([Volkov (2004)],[Volkov (2007)]). The collections of diagrams published separately could have been destroyed between the seventh and eleventh centuries, together with astronomical books.

In contrast to these early materials, the number of diagrams accompanying Song and Yuan mathematical texts is overwhelming. In the thirteenth century, *tu* were included in the texts themselves and articulated into the discourse as seen in the *Development of Pieces [of Areas]*. Some thirteenth century mathematicians inherited the ancient way of working with diagrams ([Chemla (2001)], [Volkov (2007), 445]). There are thus several ruptures in the tradition of using diagrams in the earliest sources, their commentaries and medieval treatises. Chinese mathematical diagrams cannot be considered in a uniform way because of the variety in the use and in the nature of diagrams.

In the thirteenth century, diagrams were used in several ways. They were used to illustrate shapes, form the bases of proofs of the correctness

---

[28] A large number of publications are dedicated to this domain of mathematics in ancient China. [Li (1926)], [Qian (1937)], [Guo (1982)], [Li (1982)], [Li (1990)], [Mei (1984)], [Needham and Wang (1954), vol III, pp. 21–24], [Ang (1978)], [Cullen (1996)], [Qu (1997)] among others.

of algorithms, or to supplement the processes of transformation [Chemla (2010)]. [Volkov (2007)] showed that the diagram found in the 1433 Korean reprint of Yang Hui's *Fast Methods of Multiplication and Division* of 1275 shared some common features with the processes of visualisation and transformation implied in Liu Hui's commentary. Volkov showed several diagrams representing the starting and ending configurations of a transformation [Fig. 0.2]. It seems that Yang Hui attempted to perpetuate a specific kind of approach, which Volkov characterised as follows: the diagrams referred to by Liu Hui were mainly '*focused on the representation of the general pattern of transformations of the objects rather than the structure of the objects themselves. The resulting diagrams could be called 'conceptual diagrams' and their main function was to provide descriptions of the geometrical transformations to be performed in order to justify algorithms* [Volkov (2007), 456]'.

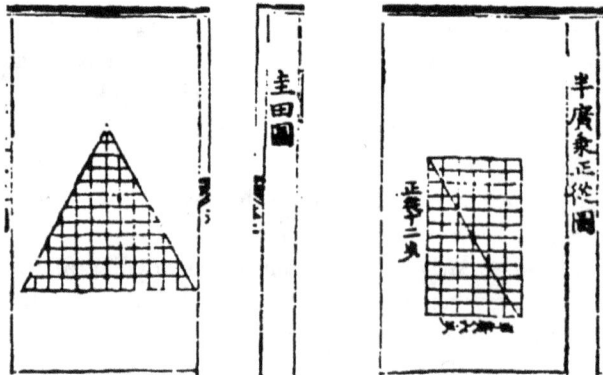

Fig. 0.2　Illustrations from the Fast Methods of Multiplication and Division, edition of 1433.

This short survey helps to situate Li Ye's diagram in the Chinese tradition. Similar to previous patterns, Li Ye's diagrams represent areas. As far as can be determined, only the *Sea Mirror of Circle Measurement* also written by Li Ye, contains a figure cut into several sections in which the vertices and line-intersections are marked by characters. Here, in the *Development of Pieces [of Areas]*, it is the areas which are marked by characters. The *Development of Pieces [of Areas]* also contains two references to colours in its discursive component. The colour red is mentioned in the discourse of two of the problems (Problem 54 and 57), but no colours or reference to colours appear in diagrams themselves.[29] It could be that diagrams were once coloured. Like the cases mentioned above, there were probably no diagrams printed in the *Collection Augmenting the Ancient [Knowledge]*. Yet, the cause of this absence might be different than the possibilities mentioned above, as will be seen later. Here, once again, diagrams correspond to the shared conception according to which they are representations of a *'general pattern of transformations'* of geometric figures. Despite the fact that there is a single drawing, there are also several layers of reading.

Indeed, diagrams are objects intended to be seen. This may sound like a tautology if it is not admitted that the verb 'to see' covers several actions or experiences. What does the visual perception produce in the present case? Is the practitioner supposed to watch and remember? Is he or she supposed to compose a mental picture? Is the practitioner supposed to draw? What is he or she supposed to understand from a drawing? What are the actions derived from this type of mute knowledge? Are they the same as what a twenty-first century reader would understand? These are some cognitive dimensions raised by diagrams [Mancosu (2005), 13–30]. The processes of visualisation by means of mental imagery are central to mathematical activity. But in what sense can mental imagery provide us with mathematical knowledge? Is visualisation more properly relegated to heuristic debates? [Giaquinto (2008)] argued that mathematical visualisation can play an epistemological role and examined the role of visual resources in the cognitive perception of abstract structures. Visual thinking in mathematics — that is, thinking with representations in visual imagination or external diagrams *Sea Mirror of Circle Measurement* has a plurality of uses. The major uses of visual thinking underpin proofs and after the construction of a proof, discovery. The role of visual thinking in proof is far from superfluous. Diagrams in the field of history of mathematics are like bones for palaeontologists. They are framework for reconstructing a whole universe of practices.

---

[29]The references to colours added inside the diagram were made the Qing dynasty editors and not by Li Ye himself. See chapter 1. III.

Part I: Text, Diagram and Equation

Despite their scarcity, diagrams play an important role in Chinese mathematics. In China, diagrams are indissociable from the history of elaboration of algebraic objects. To understand how diagrams and equations are related, recent eras offer a better point of departure than the Song–Yuan period. This chapter starts with a description of the algebraic procedure to set up quadratic equation, which is at least in part diagrammatic in origin. After the description of the procedure presented in Problem 1 of the *Development of Pieces [of Areas]*, and its general history has been established, the difference between the concept of equation as found in the procedure and modern concepts can be discussed. The origin of these differences lies in the practice of executing divisions and extracting square roots with counting rods and/or diagrams, as has been show by experts in history of Chinese mathematics. Then, a description of the geometrical procedure is introduced. This description clarifies that (1) diagrams play an argumentative role and (2) they are, ontologically, literal expressions of the mathematical object (such as the quadratic equation). An example shows how to visualise diagrams as the movement of areas and not as static lines. The chapter contains two problems, Problem 1 exemplifies the basis and Problem 21 provides a clear picture of transformations. All coloured diagrams are modern and thus artificial representation. These diagrams attempt to show, as a cartoonist might, the different cinematic moments of a movement. Their weakness lies in their modernity. Modern objects always hide contemporary concerns and purposes, probably different from Li Ye or Yang Hui's intentions. The reconstructed diagrams remain just a contemporary attempt to reconstruct ancient objects. While intended to represent movement by picturing it at separate instances, they have paradoxically stopped, like still-frames of a motion picture. However, the stop-motion animations are the best didactic tool to guide a reader through a lost 'way of seeing' mathematical object.

Chapter 1

The Celestial Source Procedure

Before tackling the question of diagrams, a detour through the famous Celestial Source procedure is necessary. This procedure stands as a lighthouse in the history of mathematics written in Chinese. More familiar than the geometrical procedure, the Celestial Source procedure is a good starting point for understanding didactic elements of the Chinese version of the quadratic equation.

1.1 Short Presentation of the Celestial Source Procedure

Much literature has been dedicated to the presentation of the Celestial Source procedure.[1] The procedure here appears as it does in Problem 1 in the *Development of Pieces [of Areas]*. This version allows a discussion of the general elements concerning its history and interpretation.

1.1.1 *Procedure*

In the *Development of Pieces [of Areas]*, the Celestial Source procedure is used to elaborate quadratic (and sometimes linear) equations with one unknown. There are no cases in which the procedure contains several unknowns or higher degrees. Each of the problems poses a question relating to field surveying as a pretext for working on a quadratic equation. The first solution to the problems in the *Development of Pieces [of Areas]* is composed of three steps, made up of the nine repetitive sentences composing the mathematical discourse.[2] These sentences give a repetitive list

---

[1][Li and Du (1987), 135–148]; [Martzloff (1987), 242–255]; [Lam (1984), 243–249]; [Qian (1964), 186–206] among others.

[2]'Mathematical discourse' refers the sentences in columns written with characters and mathematical expressions by Li Ye. Other parts of the text are composed of diagrams

of operations, which rhetorically enables the construction of mathematical expressions and implies the manipulation of counting rods on a counting surface.[3] These manipulations are never described in the text, but the reader is assumed to be acquainted with them.

As an illustration, here is a partial translation and a mathematical transcription of Problem 1 in the *Development of Pieces [of Areas]*.

Translation:[4]

> [1.1]Problem 1.
> Let us suppose there is one piece of square field, inside of which there is a circular pond. Outside the [area] occupied by water, one counts thirteen *mu* seven *fen* and a half of land. Moreover, there is no record of the [dimensions] of the inner circle and the outer square. It is said only that [the distance] from the edge of the outer field reaching the edge of the inside pond on [all] four sides is twenty *bu*.
> One asks how much are [the diameter of] the inner circle and [the side of] the outer square.
> The answer: the side of the outer field is sixty *bu*. The diameter of the inside pond is twenty *bu*.
> [1.2] The method: (1)[5] Set up one Celestial Source as the diameter of the inner pond. Adding twice the reaching *bu*
> yields $\begin{smallmatrix} 4\ 0\ tai \\ 1 \end{smallmatrix}$[6] as the side of the field.

---

and commentaries.

[3]No information concerning the counting surface exists and little information on the small bamboo counting rods survives. In August 1971, more than 30 rods of 140 mm were unearthed from tombs from the time of Emperor Xuan (73–49 BC) of the Western Han dynasty in Shanxi. In 1975, at Hubei, a bundle of rods was unearthed from a tomb of the reign of Emperor Wen (179–157 BC). The chapter entitled *Memoir on the Calendar* in 'History of the Sui Dynasty' (*Lü Li Zhi in the Sui shu*, seventh century) states that 'to calculate, one uses bamboo, two *fen* wide and three [Chinese] inches long' [Li and Du (1987), 7–8]. The counting rods gradually grew shorter, but since no later artifacts have been discovered, the range of use of the rods in the Song–Yuan Dynasties remains unknown. See also [Needham and Wang (1954), 365]; [Martzloff (1987), 194]; [Guo (1991), 26–27]; [Li and Du (1987), 6–24]; [Chemla (1982), ch. 4.3]; [Chemla (1996)] and [Chemla and Guo (2004), 15–20]; [Volkov (2001)].

[4]This translation aims to be as literal as possible. The mathematical language used by Li Ye is natural language to which he added artificial dimensions. The translation tries to render this game between natural and artificial frontiers of language. Another attempt to transcribe this artificiality can be found in [Chemla (1982)] and [Hoe (1977)]. Because of problems of legibility, another technic of translation is proposed here.

[5]The numbers in parentheses are original additions. They correspond to the numbers added to the mathematical transcription that follows.

[6]A matrix array reproduces as closely as possible the tabular settings found in the Chinese edition. The interpretation of the array is the object of the next section of this

[1.3] Augmenting this by self-multiplication yields

$$1\ 6\ 0\ 0\ tai$$
$$8\ 0$$
$$1$$

as the area of the square, which is sent to the top position.

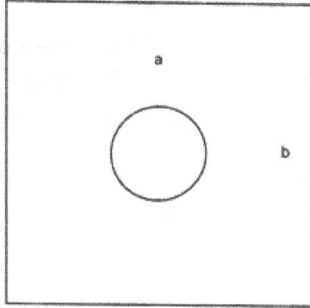

Fig. 1.1   *a*: distance to the water, 20 *bu*. *b*: side of the square field, 60 *bu*.

[1.4] Again, set up one Celestial Source as the diameter of the inner pond. This timed by itself[7] , and multiplied further by three and then divided by four, yields

$$0\qquad tai$$
$$0$$
$$0\ .\ 7\ 5$$

as the area of the pond.

[1.5] Subtracting (*jian* 減) this from the top position yields

$$1\ 6\ 0\ 0\qquad tai$$
$$8\ 0$$
$$0\ .\ 2\ 5$$

as a piece of the empty area, which is sent to the left.

[1.6] Next, place the real area. With the divisor of *mu*, making this communicating[8] this yields three thousand, three

---

chapter.

[7] Translations presented here are in part literal to reflect the terminology used in the Chinese documents. For instance, the character *cheng* 乘 Cheng 乘 to multiply is translated by 'to multiply', while the character *zhi* 之, which also designate the same operation, is translated literally as 'times'.

[8] 通 'to communicate' see [Chemla and Guo (2004), 994–998]. Here 1 *mu* = 240 *bu*. The operation of communication applies to integers *a*, *b*, *c*...and a fraction *e/f* to transform them into numbers expressed by the same unit (here *bu*). They become *fa*, *fb*, *fc*...Technical vocabulary related to fractions is not scarce in the *Development of Pieces*

hundred *bu.*
[1.7] With what is on the left, eliminating from one another

$$
\begin{array}{r}
1 \ 7 \ 0 \ 0 \\
-\ 8 \ 0 \\
-\ 0 \ . \ 2 \ 5
\end{array}
$$

(*xiang xiao* 相消) yields

Mathematical transcription:

[1.1] Let a be the distance from the middle of the side of the square to the pond, 20 *bu*; let A be the area of the square field (S) less the area of the circular pond (C), 13 *mu* 17 *fen*, or 3300 *bu* and let x be the diameter of the pond.

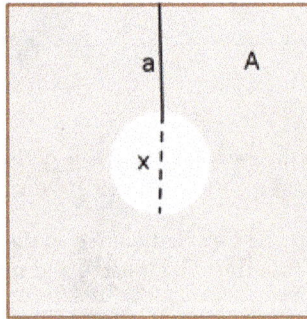

Fig. 1.2

[1.2] Side of the square $= 2a + x = 40 + x$.
[1.3] $S = (2a + x)^{22} = 4a^2 + 4ax + x^2 = 1600 + 80x + x^2$.
[1.4] $C = \frac{3}{4}x^2 = 0.75x^2$, since $\pi = 3$.[9]
[1.5] $S - C = 4a^2 + 4ax + x^2 - \frac{3}{4}x^2 = A$.
[1.6] $= 1600 + 80x + x^2 - 0.75x^2 = 1600 + 80x + 0.25x^2 = 3300$ *bu.*

[of Areas]. There is clearly an underlying fractional conception of numbers. Secondary literature has often considered numbers in the above tabular setting as decimal, but they could be linked to fractions. 5.87 could be read as $5 + 87/100$. It is possible that what Li Ye worked on with fractional conceptions. This could be the object of further discussions. See Benoit, [Chemla (1997b)], [Benoit, Chemla and Ritter (1992)].

[9]The *Development of Pieces [of Areas]* approximates the value of π as three. See chapter 6 II.1.

[1.7] The equation:

$$A - (4a^2 + 4ax + x^2 - \frac{3}{4}x^2) = 1700 - 80x - 0.25x^2 = 0.$$

From this problem, the three steps emerge as the initial operations for the general procedure:

(1) An initial mathematical expression corresponding to the area of one of the figures named in the statement of the problem is computed. This expression is understood as an expression of the first area as a polynomial in [1.2] and [1.3].
(2) A second mathematical expression corresponding to the other figure is then computed. This expression is understood as a representation of a second polynomial in [1.4]. The second expression is subtracted from the first in [1.5], or, infrequently, added (Problem 21; 23 to 30; 38; 43; 46; 63 of the *Development of Pieces [of Areas]*).
(3) The expression resulting from this operation is equal to the area given in the statement in [1.6]. The difference or sum of the two polynomials is equal to a constant, and they cancel each other out to give the final expression, which is understood as a quadratic equation in [1.7].

The same steps are found in each of the 64 problems. For each problem, first an unknown number is chosen. Then polynomials are constructed based on the condition given in the statement. Finally, the equation that governs the unknown is found. The problems are solved after setting up an equation wherein the quantity chosen for the unknown is the root. Only the positive root are ever found by means of the computation of polynomials, but the procedure of solving the equation is never given, because the practitioner is assumed to be familiar with it. Li Ye uses no term that can be rendered as 'equation', nor does he use any term to refer to the tabular settings used to represent mathematical expressions. The tabular setting for all mathematical expressions, polynomials or equations seems to be the same at first sight. However, if an 'equation' is an equality containing one or more unknowns for which the values can be determined in order to make the equality true, then the last mathematical expression satisfies the definition of an equation when it is transcribed in modern terms.

The procedure found in the *Development of Pieces [of Areas]* has a long history.

1.1.2　*Context*

In the *Development of Pieces [of Areas]*, the name *tian yuan shu* is found only in the post face by the commentator and editor, Li Rui. Li Ye did not give a name to the procedure. The name is derived from the expression Li Ye used to start each procedure: 'to set up one Celestial Source' (*li tian yuan yi* 立天元一). It was thereafter qualified as 'procedure' (*shu* 術) by the commentator.[10] What was a procedure for the Qing dynasty commentator had earlier been a mathematical object for the Song dynasty author [Pollet and Ying (2017)]. The character *yuan* is sometimes translated as 'element' and represents the unknown. The prefix *tian*, literally 'sky' or 'celestial', indicates that it is the first unknown, or the only unknown. When there are several unknowns, such as in Zhu Shijie's *Jade Mirror of the Four Sources*, they are given the following terms *tian* (sky), *di* (earth), *ren* (man) and *wu* (thing), which are said to be borrowed from Taoist philosophy. [Needham and Wang (1954), vol. III, 129] noted the disadvantage of translating *yuan* by 'element', because of the confusion with chemical elements. [Hoe (2008), 19] noticed that in philosophical texts, *yuan* is 'the source from which all the matter in the universe stems'. *Yuan* means 'the origin', 'the beginning'. However, he chose to translate it technically as 'unknown'. The preference here is to follow the ancient philosophers and Hoe by translating *yuan* as 'source'.

In the context of Chinese mathematics, there exist three early occurrences of the terms *tian yuan*. The term first appeared in Qin Jiushao's 秦九韶 *Mathematical Treatise in Nine Chapters* (*Shushu jiuzhang* 數書九章, 1247), but its usage was related to what we identify as 'indeterminate analysis' (*da yan* 大衍, lit. 'great extention') and was thus different from that of Li Ye [Libbrecht (1973), 345–346]. Qin Jiushao and Li Ye were contemporaries. However, most modern scholars agree that there is no evidence that the two mathematicians ever met. They worked independently, living far apart in rival kingdoms. If the procedure had been widespread at this time in Northern China, one would expect to find the procedure in earlier sources.[11] Strangely, nothing tangible on the *tian yuan shu* has survived

---

[10]The expression tian yuan shu was translated as 'technique of the celestial element' by [Li and Du (1987), 135] or as 'heavenly element method' by [Dauben (2007), 324]. It is not uncommon to translate the character *shu*, 術, as the technical mathematical term of 'procedure', or to use its synonym 'algorithm'. It is an effective method expressed as an ordered and finite list of well-defined instructions for calculating researched values starting from given values [Chemla and Guo (2004), 21; 986].

[11]According to [Li and Du (1987), 139], the development of the *tian yuan shu* was local. Most studies testifying or mentioning the procedure appeared in the present-day

in printed form from before the time of Li Ye. Li Ye's studies remain the earliest available evidence of this procedure. The procedure was later generalised independently by Zhu Shijie in the *Precious Mirror of Four Sources* to the 'procedure of four sources' (*si yuan shu* 四元術) which is generalised from a solution to an equation with one unknown to that of equations with at most four unknowns. These are the only three sources from which we can reconstruct the history of the procedure. Consequently, the studies of Li Ye, being the oldest remaining evidence, possess inestimable value.

In contrast, there is other evidence of the existence of this procedure prior to Li Ye's studies. In the preface by Zu Yi to Zhu Shijie's *Jade Mirror of the Four Sources*, several references report the term *tian yuan*.[12] Although there is no longer any way to access the content of the books cited, and their content is difficult to guess, the Celestial Source procedure was probably not a completely new invention at the end of thirteenth century. [Li and Du (1987), 139] placed the origin of the procedure at the beginning of the thirteenth century or earlier. [Needham and Wang (1954), vol. III, 41] believed that the procedure could be pushed 'well back into the twelfth century', as did [Qian (1964), 190–191]. [Martzloff (1987), 259], who underlined that there could be different ways of executing the procedure, wrote that 'in fact, the set of Celestial Source procedures, which has been preserved, seems to have been invented in Northern China towards the eleventh century'. [Qian (1964), 187–190] also argued that mathematicians of the Song–Yuan period could comfortably add, subtract, multiply, and divide (by integral powers of the unknown) polynomials with one unknown. Greater accuracy about the origin of the Celestial Source procedure is elusive because the thirteenth century texts have the form of a finished and mature product, already fully developed.

After its appearance in the Song dynasty but before its rediscovery by eighteenth century editors, the procedure was forgotten in China. The procedure was re-discovered in the early Qing dynasty by the mathematician Mei Juecheng 梅毂成 (1681–1763), who was acquainted with a newly im-

---

provinces of Hebei and Shanxi. This region in Northern China was the cultural and economic centre during Jin and Yuan dynasties (1115–1368).

[12] '[···] of the four unknowns of heaven, earth, man and matter (*tian di ren wu yuan* 天地人物元); there was not a single person who spoke of them. It was only later that Jiang Zhou of Pingyang compiled *Continuation of the Ancients*, Li Wen of Bolu compiled *Illuminating Courage*, Shi Xindao of Lu Quan compiled the *Bell Classic*, Liu Ruxie of Pingshui compiled *Unlocking the Ruji Method*, [···], so that later people began to know that there was a heavenly unknown' [Hoe (2008), v]. Here Jock Hoe translates *tian yuan* as 'heavenly unknown'.

ported Western algebraic method called *jie gen fang* 借根方— 'Borrowing the Root and Powers'. The method is considered a cossic form of algebra because it uses abbreviations in Chinese characters to represent different powers of the unknown. For instance, *zhen shu* 真數 is used for the constant, *gen* 根 represents the root, *ping fang* 平方 the square, and *li fang* 立方 the cube [Jami (2012), 200–210]. Anyone using this method has to remember the names of the powers to set up equations. An equation such as $x^2 - 3x + 2 = 5$ would be rendered as '一平方 − 三根 ⊥ 二 = 五', and the signs for 'plus' (⊥) and 'minus' (−) are read as '*duo*' (多, more) and '*shao*' (少, less) instead of '*jia*' (加, to add) and '*jian*' (減, to subtract) as in Song dynasty mathematics. The powers in the equation are tacitly assumed to belong to the same unknown, and no other unknown is supposed to be in the equation. All contemporary commentators were familiar with this kind of notation, yet, Mei Juecheng was the first to understand the forgotten ancient Chinese algebra Celestial Source procedure [Hoe (2008)]. Other mathematicians of the Qianlong–Jiaqing period (The Qianlong emperor 1736–1795 and his successor Jiaqing 1796–1820) struggled to prove the precedence and greater efficiency of either Chinese or Western procedures [Horng (1993a)], [Horng (1993b)]. The Celestial Source procedure was considered an ambassador of the so-called 'Chinese traditional' algebra ([Hoe (2008)], [Tian (1999)], [Liu and Dauben (1993)]). Although in the late seventeenth and early eighteenth century, the Qing emperor Kangxi caused his court to learn and digest Western science and mathematics, certain incidents late in Kangxi's reign caused distrust between the Qing court and the Jesuits, so the impulse to found a system of scientific knowledge independent of the West grew in Kangxi's court [Jami (2012)]. The academic trends later in Qianlong and Jiaqing period strengthened the press for independence. Many scholars made it their life-long calling to revive ancient Chinese learning, including Chinese mathematics, and the theory of the 'Chinese origin of Western learning' that was discussed in the seventeenth century was brought up again. However, there were also scholars, such as Li Rui's friend Wang Lai 汪萊 (1768–1813), who believed that Western mathematical methods had value and researched the basis of Western knowledge. There were fierce debates among these scholars, and also some government officials, on mathematical procedures from China and Europe, and the Celestial Source procedure was one of the methods that was used to show the efficiency and the pre-eminence of algebra from China [Horng (1993a)], [Horng (1993b)]. Li Rui who belonged to this camp, strongly believed in the Chinese roots of algebra and commented on the works of Li Ye accordingly [Pollet (2014)]. Therefore, the work of Li Ye was the object of fierce debate whether the last mathematical expression is or is not an equation [Pollet and Ying (2017)].

Reading the tabular setting for mathematical expression is the crucial element of this discussion. In the translation of Problem 1 above, this setting looks the same for polynomials and equations at first sight, but there are several ways to read expressions depending on the interpretation of tabular layouts.

1.1.3 *Writing Mathematical Expressions*

Mathematical expressions are presented on three rows with horizontal and vertical lines, which represent numbers as counting rods on a sur-

face.[13] For example,  which is transcribed as

$$\begin{array}{ll} 2\,7\,0\,0 & \textit{tai} \\ 2\,5\,2 & \\ 5\,.\,8\,7 & \end{array}$$ de-

notes $2700 + 252x + 5.87x^2$. Each row indicates a term of consecutive powers. The upper row contains the constant term, the middle row contains the coefficient of the first power of the indeterminate (x), and the lower row contains the coefficient of the second power of the indeterminate (x^2). The characters *tai* 太 or *yuan* 元 at the side of the array indicate the significance of the numbers relative to the marked position. The character *tai* on the upper row indicates the constant term. *Yuan* is used when there is no constant term; it indicates the first power. There is no specific character for the second power and there are no cubic equations in the *De-*

---

[13]Vertical and horizontal strokes are written alternatively to indicate values:

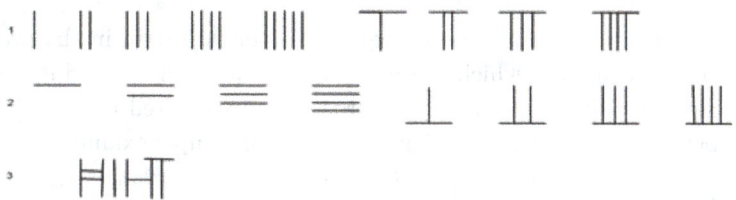

Line 1: 1, 2, 3, 4, 5, 6, 7, 8, 9 for unit, cents and ten thousands.
Line 2: 1, 2, 3, 4, 5, 6, 7, 8, 9 for tens and thousands.
Line 3: 12317

velopment of Pieces [of Areas]. The *tai* and *yuan* characters disappear in the last mathematical expression, which appears after the polynomials are 'eliminated from one another' (*xiang xiao* 相消). One of the difficulties of the text is the interpretation of the mathematical expression resulting from the operation of elimination (*xiang xiao*). Li Ye makes strict use of technical vocabulary [Chemla (2006a)], [Pollet (2012)]. The term 'to subtract' (*jian* 減) appears only when two constants or coefficients are subtracted in the construction of polynomials. The term *xiang xiao* is used only when the two polynomials are subtracted from one another to give the last mathematical expression, the one I understand as an equation.

Since the equality is never stated and the placing of rods on a tabular setting is the same for all mathematical expressions, some scholars have claimed that 'these columnar arrays of numbers do not differentiate between mere algebraic expressions and equations' [Lam (1984), 245] and 'various configurations can be regarded as either equations or polynomials' [Li and Du (1987), 138]. In fact the appearance of *tai* and *yuan* and the technical vocabulary contradicts these statements.

In Problems 38, 44, 48, 56, 59, and 60 of the *Development of Pieces [of Areas]*, Li Ye refers to different rows as 'the divisor is below and the dividend is above' (*xia fa shang shi* 下法上實). These names are also used systematically in the other procedure, the *tiao duan*, in the *Development of Pieces [of Areas]*. These are names borrowed from the procedure of division. Another expression related to division is used in Problem 23: 'the extraction of the square root by division' (*kai ping fang chu* 開平方除). The relation between division, root extraction and tabular settings, well known to historians of mathematics written in Chinese, is precisely the key to understanding Li Ye's conceptualisation of polynomial and equation.

1.2 Algorithms of Multiplication and Extraction of Square Root

This setting for a polynomial and an equation finds its origin in the procedure for root extraction, which in turn originates from the procedure for division. Several studies have already thoroughly investigated the relation of these operations. A few of the elements based on simple examples provided in the *Classic of Computation of Master Sun* (*Sunzi suanjing* 孫子算經), in which the method is described in great detail have been summarised by [Lam and Ang (2004)] [See Appendix A.2].

First, the procedure of division mirrors of that of multiplication, in which the multiplier is placed in the upper position and the multiplicand in the lower position. The latter is moved to the left according to the number of digits. The *Classic of Computation of Master Sun* prescribes multiplying the number placed below, digit after digit, by the greatest digit of the multiplier and placing the intermediate results in the middle row where they are progressively added. Similar to multiplication, whereby the operations are based on the position of the multiplier relative to the multiplicand, the operation of division is based on placing the divisor (*fa*) relative to the dividend (*shi*). The quotient (*shang*) is placed on the top. The initial position of the divisor relative to the dividend determines the position of the first digit (from the left) of the quotient.

商 *shang*
實 *shi*
法 *fa*

The characters *fa* 法, and *shi* 實, are also those used by Li Ye for the last two rows of the expression and in the second geometrical procedure. In Problems 38, 44, 48, 56, 59, and 60, *shi* is used for the constant term and *fa* for the first power in linear equations. The procedure of square root extraction has remarkable similarities with the procedures for division. Sunzi explains the method of extracting the square root with two examples in Ch. 2, Problem19–20 [Appendix A.3]. Parallelism appears with the following observations; the dividend is called *shi* and the number from which the root is extracted is called *shi*. This is the first number to be placed on the counting surface, and its digits set the values of the places for other digits on the surface. The divisor termed *fa* is moved from right to left such that its first digit from the left is placed below the first or second digit of the dividend. The quantity added at the place of *fang fa* is used like a divisor, and the quantity placed at the *shi* row is treated like a dividend. The operation returns the extraction of the root to a division [Lam and Ang (2004), 103–105].

An observation of the procedure for root extraction found in various texts shows that there is an evolution in the procedure, with the appearance of a place value notation for the equation associated with these extractions [Chemla (1994a)], [Chemla (1996)]. Chemla interpreted the parallelism between the division and the extraction of a square root thus: terms such as 'dividend' and 'divisor' for positions corresponding to the successive steps of computation have two functions. First, they allow the same list of operations to be reproduced in an iterative way. Second, they allow the extraction of the square root to be modelled on the operations for division [Chemla and Guo (2004), 327]. The same position names are used and the way of using the position during the successive operations is performed in the same way. Names and the management of positions are the key points for the correlation of the two procedures. In another publication, [Chemla (1996)] showed there is a practice that consists of exploring relations between the operations of root extraction and division; its expression is a revision of the different ways of computing and naming positions. Chemla concluded that the practice analysed in the *Nine Chapters* had been preserved until the thirteenth century, i.e. Li Ye's time.

The elaboration of the division procedure leads to a general technique that mechanically extracts the square root of a number. This method is used as a procedure but also provides the basis for the further development of procedures solving quadratic equations. The configuration necessary for conveying the meaning of these equations is inextricably expressed in the positions occupied by the rod numerals on the support. This expression justifies the use of the same divisor terms to name positions on the surface for division, extraction of the root and the terms of the equation as they are established at the same position. Once the equation is set up on the surface one just has to apply a well-known procedure to solve it. The development of the procedure of root extraction leads to the conception and solution of the equation.

It has been shown by historians that, just as the development of the method of square root extraction was based on the knowledge of division, the concept of the polynomial equation was derived from the procedure for square root extraction. The conception of the equation is procedural; it is a series of operations with two different terms — a dividend and one (or two) divisors — which is solved by the square root extraction procedure. The equality is expressed by the operation of 'eliminating one another'. However, an important premise is that only when two polynomials are equal can they 'eliminate one another'. Thus, if the question is to determine an unknown which satisfies an equality, then, in that sense, there is an equation. In fact, what we identify as equations in the representation of tabular settings is the opposition of a 'dividend' and other terms. This peculiarity of the concept of an equation is due to the essential role played by the counting surface and the way of assembling different procedures for division and root extraction.

In Li Ye's studies, the last mathematical expression resulting from *xiang xiao* (eliminating from one another) never contains a *tai* 太, or *yuan* 元

character, i.e., the character represented in the polynomials is absent from the equation. This is because once the procedure of setting up the equation is completed the last expression will not be the object of further operations [Chemla (1982), note (a) in Ch. 8.3]. The number of ranks is the only pertinent information for extracting the positive root. Having obtained the equation after cancellation, the marks on the right can be forgotten. Thus, for Li Ye, there is a difference between the mathematical expressions for polynomials and equations. The configuration can produce either an equation or a polynomial, but by adding the character *tai* or *yuan* to the column, Li Ye makes what we call a 'polynomial'. The absence of a sign is precisely the mark of the equation.

In other words, the procedure that had been presented to us and was later called the Celestial Source procedure by Qing dynasty commentators is a sequence of operations with an object named *tian yuan* for Li Ye. This sequence gave a shape to a problem in the form of 'an equation', or rather, in the form of a square root extraction applied to polynomials. In addition, this operation would be the next procedure applied in order to solve the problem. For Li Ye, *tian yuan* is the name of a component in a multi-step procedure preceding the solution of a problem. This procedure interprets the geometrical data through polynomials and then transforms the problem into a new object we now can call an equation. For Li Ye, there were several procedures which merged into an over-arching procedure, which ends with an equation which is also an operation.

1.3 The Geometrical Precursors

The equation operation has another aspect. Besides the numeric dimension linked to the algorithm of division and the manipulation of counting rods, there is a diagrammatic dimension. An equation is also established geometrically. In the introduction, it was noted that there are no known precursors to the Section of Pieces [of Areas] procedure. Relying on Li Ye and Yang Hui's testimonies, one can trace the procedure only back to the eleventh century. However, this procedure has features in common with the geometrical proof of the extraction of square roots given by Liu Hui in his commentary to the *Nine Chapters*. Problem 14 of chapter 4 of the *Nine Chapters* contains a commentary on the extraction of a square root using geometry. The concept of square root extraction was based on geometrical considerations, as is evidenced by some of the names of the positions on the counting support [Lam (1977)], [Martzloff (1987), 222–223]. Although no diagrams appear in the *Nine Chapters* or its commentaries, the commentary by Liu Hui described the existence of a coloured diagram, which he introduced to discuss a correction of the procedure of square root extraction [Chemla (2010)], [Chemla and Guo (2004), 323].

The oldest diagram in a text on showing the derivation of square root extraction appears in Yang Hui's *detailed analysis of the mathematical methods in the 'Nine Chapters'* (*Xiangjie jiu zhang suanfa* 詳解九章算法, 1261).[14] The diagram illustrates a problem from the *Nine Chapters*, which consists of finding the square root of 71,824. In addition, Yang Hui explained the method and showed the different stages on the counting support [Lam (1977), 96–97] [Chemla and Guo (2004), 323].

Consider the following reconstructed diagram: [Fig. 1.3]

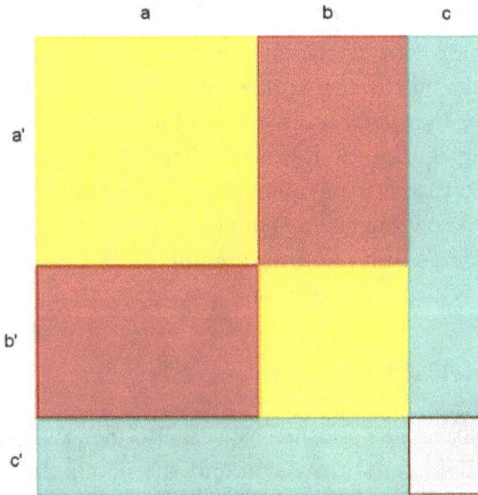

Fig. 1.3

With this diagram, the reader can follow the steps of the root extraction. If the reader wants to find the root of 234,256 (derived from *Classic of Computation of Master Sun*'s example: 234,567 minus the remainder 311), the correlation of the geometrical concept and its arithmetization can be

---

[14]It is preserved in the *Great Canon of Yongle*, Ch.16, 344. p.8a.

identified on the counting support. Rephrasing and walking in Lam Lay-yong steps [Lam and Ang (2004), 108–109], the extraction of root follows:

The total area of the square is $234,256$. The segment a represents $a.10^n$, which is the length in hundreds (i.e. 400). The segment b represents $b.10^{n-1}$, which is the length in tens, (i.e. 80) and c represents the length in units (i.e. 4). The derivation entails three stages: (1) the removal of the yellow square, (2) the excision of the red and yellow gnomon and (3) the elimination of the blue gnomon. These three steps are presented in relation to the algorithm for the extraction of square roots.

(1) When the length a is obtained (Step 3), a similar length is identified (a') on the *fang fa* ('square divisor') (Step 4), and the yellow square area is removed (Step 5). [Appendix A.3].

(2) The doubling of a' (Step 6) implies two similar lengths of the red rectangles less the side of the little yellow square. When b is obtained (Step 8), a similar length is identified (b') and placed on *lian fa* ('side divisor') (Step 9). The two red rectangles and the little yellow square are removed from the remaining area (Step 10. Appendix A.3).

(3) The doubling of b' implies two similar lengths (sides of the little yellow square). When these two lengths are adjunct to the side of the big yellow square, the side of the blue gnomon is produced, less a small square (Step 11). When the length c is obtained (Step 13), a similar length is identified (c') and placed at *yu fa* ('corner divisor') (Step 14). This last gnomon is the remainder of the area to be removed (Step 15. Appendix A.3).

There are several points in common with the *Development of Pieces [of Areas]*. First, although there are no colours in the available editions, red is mentioned twice in the discourse by Li Ye about the 'meaning' of Problems 54 and 57. The two sentences in question are: '*Inside the area of the expanded pond, one sets four pieces of red area*' and, '*the eight empty [areas] are recorded in red*'.[15] Both times, the colour red is applied to areas which are removed later.

Second, in the extraction of the root presented by Yang Hui, the dividend is visualised as a basic area from which one progressively removes the products of the length, the root being only momentarily considered. Each of the products are also visualised on the figure by pieces of area which have to be removed. The process is complete when the basic area is ex-

---

[15]'所展池積內, 將四段紅積' and '八處以紅誌之者'. The character 'red' which appears eight times in the diagram was added by Li Rui.

hausted. This practice of decomposing and recomposing areas also informs the reading of the diagrams of the *Development of Pieces [of Areas]*. This practice verifies the validity of the provenance of the terms of the equation. A similar practice is mentioned by Yang Hui, who in turn connects this practice with that of Liu Hui. [Horiuchi (2000), 248] proposed the following hypothesis concerning the configuration of the diagrams in *Yang Hui's Methods of Computation*: it may be that the diagrams not only translate the condition of the problem and the origin of the equation, but it could also be that the figure is a based on which one was used to perform the extraction of the root. It could also be that the diagram shows the best possible geometrical configuration to illustrate the procedure of root extraction for each type of quadratic equation. In the *Development of Pieces [of Areas]*, Li Ye focuses on establishing equations and not on their solution. The importance of this difference will be established later.

Another remarkable point concerns equations. In the ninth of the *Nine Chapters*, a new object is introduced: the quadratic equation. In the commentary by Liu Hui, a figure can be reconstituted which corresponds to the equation [Chemla and Guo (2004), 690] from a rectangle whose area is known, inside of which there is one square whose area is x^2 and a second rectangle with one of its dimensions known. The first rectangle is called a piece of 'area at the outside of the corner.' Here, Liu Hui makes a fundamental reference to the extraction of square roots. Indeed, after removing the first square (see Step 1), Liu Hui's algorithm structures the remaining gnomon by opposing the square in the corner and the two equal rectangles. The figure for the quadratic equation presents a square comparable to the square in the corner in the extraction of the root, as well as two rectangles corresponding to each other. Chemla showed that in both cases, the general area is known, and the purpose is to determine that of a square in the corner.

Quadratic equations as numerical operations, are modelled on the algorithm of the extraction of square roots. The terms of the equation take their meanings from the references to the procedure of manipulations on the counting support for extracting the square root. In his commentary on the *Nine Chapters* mentioned above, Liu Hui established a link between a diagram illustrating the algorithm for the extraction of the root and a fundamental figure of the equations. The conception of the equation in the Celestial Source procedure finds its origin in the tabular procedure for root extraction. In the procedure of the Section of Pieces [of Areas], the relation of the equation to the root extraction procedure is the same. The diagrammatic concept of the equations is modelled on the diagrammatic demonstration of the correctness of the procedure for root extraction.

The geometric connection between the procedures of the Celestial Source and the Section of Pieces [of Areas] appears throughout the mathematical discourse of the *Development of Pieces [of Areas]*. The Celestial Source procedure is not a procedure abstracted from signification. A translation into geometrical terms follows each of the tabular settings. The discourse is constructed following the constraint that the polynomial expressions of each of the geometrical intermediate steps must be able to be computed. These steps are expressed with the character *wei* 為 which is translated here as 'as' or as 'to make'. This character is used as a pivot for connecting the Celestial Source to Section of Pieces [of Areas] procedures.

In the Celestial Source procedure, *wei* translates polynomials into geometrical concepts.

This is exemplified in the following passage (Problem 5): '*Set up again the Celestial Source, the circumference of the circle. Self-multiplying this yields* $\begin{smallmatrix} 0 \\ 1 \end{smallmatrix}$ *yuan* *as twelve pieces of the area of the circle pond*'.[16]

Each of the results obtained through the procedure is interpreted in geometrical terms, or 'pieces [of area]' *duan* 段. Polynomials are representations of geometrical quantities with one or two known quantities and one unknown quantity. The equation is thus established at the end of the geometrical interpretations of the given problem.

In the Sections of Pieces [of Areas], the character *wei* translates geometrical figures into expressions of arithmetical coefficients and positions on a support. This correspondence with geometric figures is exemplified in the following (Problem 5): '*From forty-eight pieces of the area of the field, one subtracts three pieces of the square of bu of the difference to make (wei) the dividend. Six times the difference of the bu makes (wei) the joint* [⋯]'[17]

The two procedures are expressed in such a way that each reference the other, which makes it is difficult to interpret the procedures as being autonomous. The mode of expression explains why they are being juxtaposed and confirms the geometrical origin of the procedure of the Celestial Source, and thus, the antecedence of the procedure of Section of Pieces [of Areas].

---

[16]再立天元圓周. 以自之 為十二段圓池積.
[17]四十八段田積內減三段不及步冪為實. 六之不及為從.

This shows that equation has in fact two aspects. In addition to being a numerical operation, it has a diagrammatic aspect, each part of the diagram being the support of part of the working of one equations. In fact, an equation was established and stated geometrically, in the form of a rectangle having a given area (the dividend) and one or two rectangles, i.e., the square of the unknown and another rectangle, a side of which was the unknown and the other side of which corresponds to the adjunct divisor. This structure explains how the two facets of the equation — the numerical aspect and the geometrical — were related to each other. [Chemla (2016)] established that the two facets of the equation found in the *Nine Chapters* show a dramatic shift reflected in the works of Li Ye. The geometrical reading of the equation has been entirely erased from the Celestial Source procedure, because the work on the equation relies entirely on numerical tabular settings. This shift is characterised by the disappearance of the diagrammatic aspect and would have remained invisible had its earlier significance had not been brought to light. The *Development of Pieces of Areas* plays the role of missing link in this reconstitution.

Chapter 2

The Procedure of the Development of Pieces of Areas: The Example of Problem 1

2.1 Diagram and Text

2.1.1 *Meaning of Yanduan*

Whereas copious literature treats the Celestial Source procedure, scant secondary literature considers the Section of Pieces [of Areas] procedure in Western languages. The English translation of [Li and Du (1987)] did not mention this procedure and [Martzloff (1987), 143] dedicated a solitary note to this procedure: *'The term yanduan is difficult to translate. It means an algebraic technique which depends both on computation and on geometric figures'*. [Lam (1984)] provided a translation and mathematical transcription of Problem 8.[1] Lam Lay-Yong hoped that *'the analysis of this problem gives the reader a general notion'* and surmised that the procedure of the Celestial Source is a *'general technique'*, while that of sections of areas is *'abstruse and has no fixed approach'* [Lam (1984), 249]. According to Lam Lay-Yong, the approach depends on the specific data of each problem. Here, the question of generality of the procedure arises because it is possible to view this procedure as another way of expression of generality.

The term *duan* 段 appears in two expressions: *yan duan* 演段, and *tiao duan* 條段. In the *Development of Pieces [of Areas]*, the expression *yan duan* appears only in the title, while the expression *tiao duan* appears in the problems. The expression *tiao duan* appears in the preface by Li Ye and in the first sentence of the second procedure of each problem: 'one looks for this (i.e. the unknown) according to [the procedure of] the Section of Pieces [of Areas]' (*yi tiao duan qiu zhi* 依條段求之). In *Yang Hui's Methods of Computation*, the expression *tiao duan* does not occur, but the expression

---

[1]This translation was reproduced in [Dauben (2007), 328]. This Problem 8 is also presented in [Kong (1988)], [Kong (1996)], [Kong (1999)].

yan duan appears in the titles of problems 46 to 53 and 62 to 64: 'the development of pieces:···' (*yan duan yue* 演段曰).[2] In *Yang Hui's Methods of Computation*, the expression appears in problems related to rectangular fields and the setting up of quadratic equations. Although the first eight problems of the chapter also deal with rectangular fields, the procedure focuses on the ways to solve quadratic equations and not on methods for composing the equations. Problems 54 to 61 concern circles, annuli and trapezoids but are not titled *yan duan*.[3] In the *Development of Pieces [of Areas]*, the procedure concerns rectangles, squares and circles.[4]

The term *duan* means 'piece', 'portion', 'part' or 'section'. Horiuchi suggested that this term is related to areas and translated *yan duan* as 'development of pieces of areas'.[5] The word *duan*, however, is not a synonym for the contemporary term for area, *ji* 積 (area as product) or for the term *mi* 冪 (surface) from the computation of area and volume in older texts.[6] In many sentences on the Celestial Source procedure in the *Development of Pieces [of Areas]*, *duan* is used as a numeral classifier[7] for reckoning areas involved in a computation. For example, Problem 13, 以三之得 $[2700 + 352x + 4.87x^2]$ 為四段外圓積 may be translated as 'Tripling this yields $[2700 + 352x + 4.87x^2]$ as four **pieces** of areas of the outer circle'. Following Horiuchi's translation, *duan* may be read as 'piece' and *yan* as 'development'. *Tiao* is also a numeral classifier of 'pieces'. This character may be translated as a synonym for *duan*, 'section'. Therefore, the object that is cut into pieces has been consistently supplied in brackets. Thus, *tiao duan* is translated here as the 'Section of Pieces [of Areas]' and *yan*

---

[2]According to the numbering by [Lam (1977)]. In [Guo (1996)], the same problems are numbered 9 to 20 and 26 to 28.

[3]The fact that *yan duan* appears only in the title (*Development of Pieces [of Areas]*) but Li Ye prefers *tiao duan* in the discourse suggests synonymy between the two expressions. No hypothesis explains the evolution of these two expressions, nor does a reason exist why the expression is not a category of problem in *Yang Hui's Methods of Computation*.

[4]Three problems of *Development of Pieces [of Areas]* present no solution found from the Section of Pieces [of Areas] procedure. These problems concern the irregular quadrilateral and three figures inscribed in each other. The names of the terms in the polynomials also differ. At the end of these problems, a procedure named 'Old Method' (*jiu fa* 舊 法) is found instead of 'Old Procedure'. These cases are not investigated here, but it is interesting to note that, in *Yang Hui's Methods of Computation*, problems concerning irregular quadrilaterals and annuli are not titled '*yan duan*' either.

[5][Horiuchi (2000), 245]: 'développement des pièces d'aires'. [Lam (1977), 86, 118] translated *yan duan* as 'procedure'.

[6]See [Chemla and Guo (2004), 932–933 and 959–961].

[7]A classifier, sometimes called a counter word, is a word or morpheme which accompanies nouns and 'classifies' the noun according to the type of its referent.

duan as the '*Development of Pieces [of Areas]*.'

2.1.2 Different Types of Diagram

Each of the sixty-four problems starts with the statement of a question and its answers, as illustrated by the first diagram. The text of the procedure of Section of Pieces [of Areas] takes the shape of two small paragraphs accompanied with second diagram. There are few exceptions to this general format of text. Some problems contain a third diagram in the Section of Pieces [of Areas] (Problems 21 and 64), and others omit any caption (Problem 6). One problem is presented without a specific diagram for the Section of Pieces [of Areas] description and without the usual list of operations, although the 'meaning' is still given (Problem45). For the twenty-three of the problems with a description of the Old Procedure, three are presented without a description of the procedure of Section of Pieces [of Areas] (Problem 44, 59 and 60). The Problem 22 includes a specific diagram for the Old Procedure. A systematic study will reveal some clues toward an understanding of these peculiarities and how the diagrams are articulated with the lists of operations and the 'meanings'.

The discursive part never directly mentions 'diagram', *tu*, 圖 but it implies an important set of non-discursive practices with diagrams. The diagrams are supposed to be drawn by the reader. Thus, there three types of diagram: those edited into the statement of the problem, those edited into the descriptions of the Section of Pieces [of Areas] procedure by Li Ye himself, as he indicated in his preface, and the drawings made by the practitioner.

As printed elements, diagrams are part of the text but are not discursive elements. The discourse written by Li Ye almost never refers to diagrams. The character *tu* 圖, 'diagram', appears very few times in the *Development of Pieces [of Areas]*, whereas the character *shi* 式 ,'pattern', 'configuration', is used more than twenty times in both procedures. In fact, the character *tu* appears only five times in commentaries and prefaces by Li Ye. The word appears in Problems 45 and 61 to recommend carefully drawing two of the diagrams of the Sections of Pieces [of Areas]; in Problem 22 to indicate that one of diagrams is the diagram for the Old Procedure and in Problem 64 to mention that the problem atypically contains three diagrams. Each time, this character refers to visual artefact inserted inside the text, that is to say, diagrams. The character *tu* never appears in the mathematical discourse on the Section of Pieces of Areas but only in commentaries composed by

Li Ye.

It seems that *tu* and *shi* refer to different objects. *Tu* refers to a specific category for technical images. As [Schäfer (2011), 141] and [Bray (2007), 4] note, *tu* is functional: "the images convey skilled, specialist knowledge and complex meanings with more or less succinct methods [···] Diagrams give realistic impressions of 'actual' object as well as representational, conceptual portrayals of complex structures". Schäfer noted that they are templates for action. Diagrams, as technical images, have representational and documentary functions. In this light, *tu* extend beyond illustrations. Here, they will be used in an active role to help establish proofs of mathematical theorems. In contrast, the other essential visual tool, *shi*, names the configuration of the rods on the counting support reproduced inside the written column and also in the figure drawn by the practitioner in the Section of Pieces [of Areas] procedure. The character *shi* refers to pictorial representation of material objects involved construction. This could confirm an absence of diagrams in the materials used by Li Ye for composing his treatise. The disparity of these two characters is a first step toward questioning the nature of the authorship of the treatise and moments when it was composed.

The word *tu* appears only once in the Celestial Source procedure, and surprisingly, it does not concern any geometrical diagrams. The fifth occurrence of the term *tu* is in Problem 63 [Fig. 2.1], though, particularly interesting.

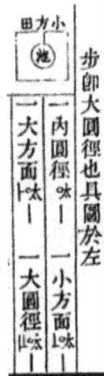

Fig. 2.1   Problem 63

I translate this part as

A diagram is provided on the left:

One diameter of the inside circle: $\frac{0\ tai}{1}$

One side of the small square: $\frac{60\ tai}{1}$

One side of the big square: $\frac{110\ tai}{1}$

One diameter of the big circle: $\frac{160\ tai}{1}$

In this example, four polynomials are presented in a separate list of two columns rather than inserted inside a sentence [Fig. 2.2]. Moreover, this time, the author names this configuration *tu*, 'diagram'.

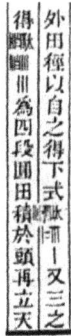

Fig. 2.2   Problem 11

[Chemla (1996)] showed that there were different practices of representation, and that the transcription of tabular settings was not uniform in the Song–Yuan period. Li Ye distinguished himself from contemporary mathematicians by the way he elaborated a transfer of the mathematical activity to the paper. That is, he developed a way to represent polynomials and equations specifically for the written work and different from the ancient practice of manipulating counting rods. When Li Ye wrote mathematical expressions according to this method, he has been interpreted as creating symbol for the object he discussed. Chemla has shown that different texts manifest different positional notations from the Han to the Yuan Dynasties. Matrix arrays are continuously used, but there are variations in their transcriptions. This variation shows an evolution toward autonomy of work on paper. Despite these variations, the organisation of the data is remarkably stable and the management of the operations on the support is led by strict imperatives. This practice testifies of a transition movement of the mathematical activity from the support to a paper based work and of development of symbolisation.

To understand the peculiarity of Li Ye's methods of writing polynomials, refer to other contemporary mathematicians. For example, Qin Jiushao seems closer to a pictorial configuration. On the contrary to Li Ye, Qin Jiushao inserts the diverse state of the support as an illustration. The discourse and its illustrations are discriminated, with the discourse sometimes being a caption to an illustration. We also notice that Qin Jiushao refers to the tabular setting using the character 'diagram' (*tu* 圖) [Fig. 2.3]. The configuration of numbers introduced by Li Ye portrays some states of this support, yet we cannot consider this as a pure transcript of the different steps of manipulating the rods. The configuration, as Li Ye presented it, is not a picture of the support; it is a step in putting down the symbolisation.

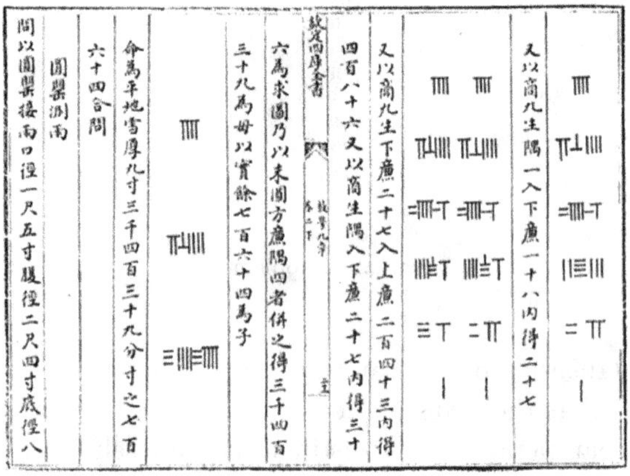

Fig. 2.3   Ch. 2. p. 21

In the available editions of the *Development of Pieces [of Areas]* and in the *Sea Mirror of Circle Measurement*, [Chemla (1996)] observed that the mathematical expressions are always written in the space of the column containing the text, just like a part of a sentence. They are introduced by the character 'to yield' (*de* 得) and are then interpreted with the character, 'as' (*wei* 為). They are not represented as independent drawings [Fig. 2.2]. There is continuity in the written text between the discourse and the configuration (*shi* 式) of numbers. Li Ye integrates the configuration into the written text as if it were a simple number, whereas the configuration itself extends the sentence as across the column. There are no such relations like picture/caption; therefore, configurations cannot be considered illustrations either. Only the case of Problem 63 presents a different format in which the layout is spread across two rows and over two columns, but this configuration has the specific appellation '*tu*'. Li Ye clearly makes a distinction between a configuration as a visual artefact and a configuration as part of the written discourse. The character *tu* refers to the first. This relation between the diagrams and the discourse is also problematic for geometrical diagrams.

A diagram follows each statement of the problem. Some data from the statements are reported inside the diagram. This kind of diagram illustrates and summarises the data for the problem, generally naming the square field and the pond (if this is the case) and indicating the distances that are already known. Sometimes the expected results are already written down. These notations are not systematic and vary from one problem to the other.

Close inspection reveals that the majority of these diagrams contain only the data for one distance, and this distance is named according to a system of abbreviation. Areas and segments are given in the same unit without differentiation, the *bu* 步. Nothing corresponds to modern square units for areas. These values are always expressed in natural language in the statement, and thereafter referred to through abbreviations. For example, in the first problem '[*the distance*] *from the edge of the outer field reaching the edge of the inside pond is twenty bu for each side*' 從外田楞至內池楞四邊二十步 is reduced to '*the reaching bu*' 至步. The abbreviation is reported as a caption in the diagram and is used to name the segment in question in the different procedures.

The distance drawn or within a caption in the diagram is always the one used in the construction of the first polynomial by the two procedures. As noted in [Table 2.1], many diagrams contain no caption for the data at all, but the distance given in the statement is always drawn. The diagrams with answers satisfy only one of the several questions being asked. The answer given will be used to deduce the other answers. Curiously, a small number of diagrams contain data which are neither in the statement nor in the answer. These concern only perimeters or circumferences for which the side or diameter is given instead. Perimeters and circumferences can be deduced from the side and the diameter given in the diagram. From these observations, the data given in the diagram statement forms the basis from which the other data are deduced, and these quantities are used to set up the algorithm.

Table 2.1: Types of data contained in diagrams in the statement

	Problem	Total
One of the answers	1; 7; 9; 10; 11a; 13; 14; 19; 20; 61; 62	11
Other data not named in statement	5; 6; 7; 9; 16; 17; 18	7
No caption of the data at all	8; 15; 21; 23; 25; 26; 27; 28; 29; 30; 32; 33; 38; 44; 46; 47; 49; 55; 57; 58; 59; 60; 63	23
Distance named in the statement	2; 3; 4; 7; 9; 11a; 12; 13; 14; 22; 24; 31; 34; 35; 36; 37; 39; 40; 41; 42; 43; 45; 48; 50; 51; 52; 53; 54; 56; 61; 62; 64	32

The quantity written or represented in diagrams is always the constant that is added to or subtracted from the unknown in the Celestial Source procedure in order to set up the first polynomial. The diagram not only illustrates and summarises the statement; it also plays another role. By representing other objects required by the procedures, it represents the first step of the algorithm. So, the question may be posed: Do these first diagrams only illustrate the data of the statement or are they linked with execution of the procedures? If the diagrams are linked with the procedures, how are they connected to the second diagrams presented in the descriptions of the Section of Pieces [of Areas]?

To understand the role of the diagram, one has also to keep in mind that in the descriptions of the Section of Pieces [of Areas] procedure never explicitly states the equation. The equality is never mentioned, nor is the concept of an unknown; moreover, there is no symbol to express the sign of the coefficients.

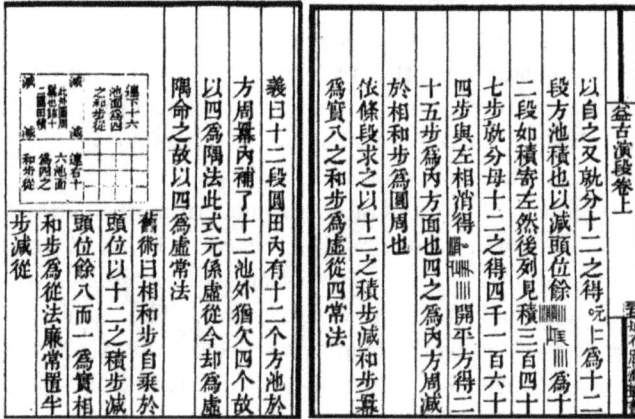

Fig. 2.4 *Development of Pieces [of Areas]*, pb18, republished in 1798 by Li Rui.

The Celestial Source procedure in the *Development of Pieces [of Areas]* contains many representations of quadratic equations. As seen before, the equation is presented on three rows with numbers represented by counting rods on a counting surface. The descriptions of the Section of Pieces [of Areas] procedure does not use such a tabular representation, but nonetheless mention rows.

For example, consider the sentence in Problem 8:

'One looks for this according to the Section of Pieces [of Areas]. From the square of the *bu of the sum* (和步), one subtracts sixteen times the real area (見積) to make the dividend. Six times the *bu of the sum* makes the adjunct. Three *bu* is the constant divisor'.[8]

The constant is placed on the row called the 'dividend'; the first-order indeterminate term on the 'adjunct' row, and the second-order indeterminate term on the 'constant divisor' or sometimes 'corner' row. These words are abstract, technical names for the coefficients and the names of positions on a counting surface on which counting rods are manipulated, as has been

---

[8] '依條段求之和步幂内減十六之見積為實. 六之和步為從. 三步常法' In the Section of Pieces [of Areas] procedure, numbers are expressed rhetorically, and thus translated literally. In the Celestial Source procedure, numbers are expressed by counting rod-based notation. Both, *jian ji* 見積 and, *zhen ji* 真積 (found in other problems) are translated as 'real area', because the area discussed in the problem is expressed in terms of tangible constants, as opposed to quantities expressed by unknown variables. This expression is distinct from 'equal area' (*ru ji* 如積) and 'empty area'(*xu ji* 虛積) which express the same area in polynomials.

made clear by a number of historians. What are here called 'terms' are the entries on different rows of a counting surface.

The '*bu* of the sum' and the 'real area' of the example quoted above are quantities known from the statement of the problem. The '*bu* of the sum' denotes the sum of a perimeter and a circumference, and the 'real area' is the area of the circle subtracted from the area of the square. Both are constants measured in units of *bu*. In modern mathematical notation, the first sentence would read $a^2 - 16A$, where a is the '*bu* of the sum' and A is the 'real area'.[9] The second sentence corresponds to $6ax$, and the final sentence would be represented as $3x^2$. The equation would be $a^2 - 16A = 6ax + 3x^2$, which relates two coefficients of the equation to the known values of the '*bu* of the sum' and the 'real area' with a directly stated third term. The sentence, however, also relates which row of the counting surface the quantities occupy (dividend, adjunct or constant divisor). In fact, in this procedure, geometrical diagrams have replaced tabular representations. This replacement, together with an understanding of the manipulation of counting rods on the counting surface given in the texts, enables insights into the changing conceptualisation of these equations. As stated before, it is not clear how the data given in the statement of the problem is transformed into an equation. This is where an understanding of the diagram and the manipulations on the counting surface become important.

After the statement of the question and the answer, the problem presents a diagram containing captions, accompanied by a paragraph entitled 'meaning'. The 'meaning' links the diagram to the preceding list of operations. The 'meaning' contains the operations to be performed on the counting surface and advice on how to proceed with different pieces of the diagram, or how to draw it. This discourse is not narrative: it neither describes nor explains the successive steps of the algorithm. Only a few sentences give 'hints' to some particular steps of the algorithm. Li Ye mentioned only those details he deemed necessary, but the algorithm is never entirely described. For instance, in the two cases of Problems 3 and 43, the 'meaning' concern some supplementary operations which were not listed in the first sentence of the description of the Section of Pieces [of Areas]. Li Ye gave a few instructions and showed how an action should be done, or how the exercise should take place. Instead of giving definition or ready-made interpretation, he directly gave the usages of signs. The 'meaning'

---

[9] In this study, the measured area is consistently reported in *bu*, 步, given in the statement as A for areas, and a for distances.

actually explains nothing. Its purpose is not to state the reason for a rule but to indicate its application and thereby give a succinct operational description. The 'meaning' comments on the function of the diagram rather than describes or enumerates a complete list of operations. The diagram is therefore central to the 'meaning' and it communicates something not enunciated by the 'meaning'. The reader must complement the 'meaning' with the diagram.

For comparison, *Yang Hui's Methods of Computation*, like the *Development of Pieces [of Areas]*, starts each problem with a statement and its answer. No diagrams are used to illustrate this part; then, a sentence titled 'procedure' (*shu* 術) explains how to compute the coefficients in vocabulary similar to that of the *Development of Pieces [of Areas]* (dividend, adjunct, corner). The term 'constant divisor' never appears. This sentence is followed by a paragraph titled *yan duan*. This paragraph functions like the 'meaning' in the *Development of Pieces [of Areas]* and refers to a diagram and its caption (Problems 49, 63 and 64 are not accompanied by diagrams). The rest of the problem begins with the title *cao* 草, 'working' [Lam (1977), 87] and solves the quadratic equation. This solution is accompanied by a diagram in problems 46 and 47.

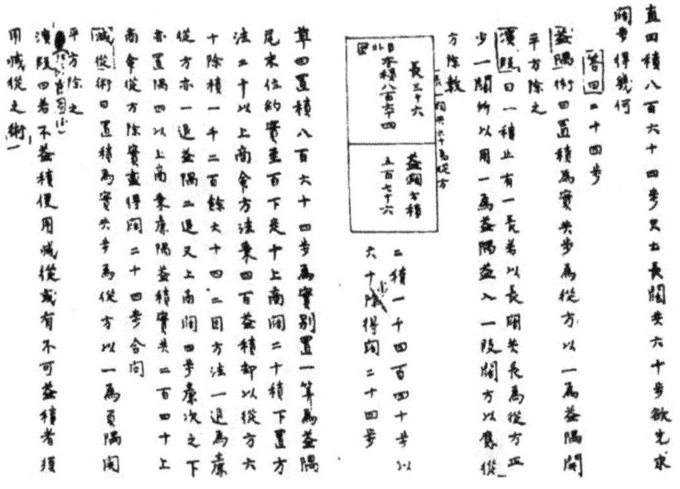

Fig. 2.5 *Yang Hui's Methods of Computation*, republished in Guo Shuchun 1993.

In the *Development of Pieces [of Areas]*, the procedure is to set up an equation but the means of solution are never given. In *Yang Hui's Methods of Computation*, all the problems of the second chapter focus on the solution and the various procedures for extracting the roots. Both authors distinguish two phases of solving a problem, the creation of the equation and the extraction of its roots. Li Ye focused on the former; Yang Hui emphasised the latter.

2.2 Diagram as Equation

Here, a basic problem, Problem 1, explains the rudiments of the Section of Pieces [of Areas] procedure.

2.2.1 *Translation*

[1.1] Problem 1.

Let us suppose there is one piece of square field, inside which there is a circular pond. Outside the [area] occupied by water, one counts thirteen *mu* seven *fen* and a half of land. Moreover, there is no record of the [dimensions] of the inner circle and the outer square. It is said only that [the distance] from the edge of the outer field reaching the edge of the inside pond on [all] four sides is twenty *bu*. One asks how much are [the diameter of] the inner circle and [the side of] the outer square.

The answer: the side of the outer field is sixty *bu*. The diameter of the inside pond is twenty *bu*.

[1.8] Look for this according to the Section of Pieces [of Areas]. From the real area (真積), four pieces of the square of the reaching *bu* (四段至步冪) are subtracted to make the dividend (實). Four times the reaching *bu* (四之至步) makes the adjunct (從). Two *fen* and a half is the constant divisor (常法).

[1.9] The meaning: To subtract (內減) four pieces of the square of the reaching *bu* from the real area is to subtract (減去) the four corners. [Taking] two *fen* and a half as the constant divisor, [it] is [found] that for each *bu* of the inside [part] full [of water], seven *fen* and a half are [taken] off [and] outside there are two *fen* and a half.

2.2.2 *Mathematical Transcription*

[1.1] Let a be the distance from the middle of the side of the square to the pond, 20 *bu*; let A be the area of the square field (S) less the area of the circular pond (C), 13 *mu* 17 *fen*, or 3300 *bu*; and x be the diameter of the pond. Find x, the diameter. See [Fig. 2.8]

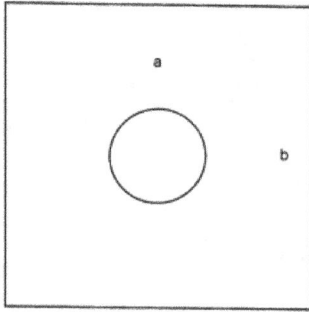

Fig. 2.6 *a*: 至水二十步. The *bu* reaching the water are twenty. *b*: 方田六十步. The side of the field is sixty *bu*.

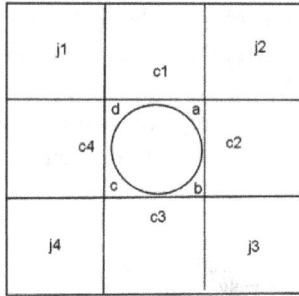

Fig. 2.7 j1–4: subtract; c1–4: adjunct; abcd: two *fen* five *li*; J1–4: 減; C1–4: 從. Abcd: 二分五厘.

Because the paragraph [1.8] opens the description of the Section of Pieces [of Areas] procedure, the obvious interpretation in modern terms is that A is 'the real area', namely, the area (that of the field minus the pond) given in the original statement and a is 'the reaching *bu*', that is, the distance expressed in units of *bu* starting from the middle of the side of the square and reaching the diameter of the pond. If $A - 4a^2$ is con-

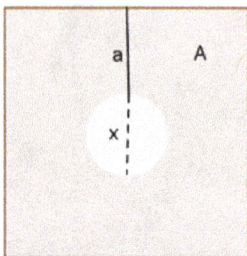

Fig. 2.8

sidered the 'dividend' (that is, the constant term), then $4ax$ is the 'adjunct divisor' (that is to say the term in x), $0.25x^2$ is the 'constant divisor' (that is to say the term in the x^2) and the equation would be given by: $A - 4a^2 = 4ax + 0.25x^2$.

Yet this interpretation overlooks the fact that the equality and the unknown are never stated in the procedure. Using a symbolic transcription of the statement of the problem into the terms A and a, the coefficients are tabulated. This table, however, represents a modern creation that is not in the text. There are only three separate elements, which may be artificially schematised in [Table 2.2].

Table 2.2

$A - 4a^2$	dividend
$4a$	adjunct
0.25	constant divisor

In [1.8], the means of computing the 'dividend' and the 'adjunct' are given, but the 'constant divisor' is stated directly as an answer and, from the first sentence, the origins of the coefficients are unknown. This sentence presents the coefficients of the equation as the final results without justification. On the basis of this sentence only, it is difficult to understand how and why to transform the data given in the statement into the equation. This example illustrates the difficulty of associating an equation, as it is defined by modern mathematics with this procedure.

In each of the problems of the *Development of Pieces [of Areas]*, the procedure is interpreted as leading to tabular settings, and the three coefficients are always listed in the same way: 'the dividend', 'the adjunct' and 'the constant divisor' or the 'corner'. The rank of the dividend on the support is filled with the quantity which will be diminished or augmented until its exhaustion. Once the numbers are set on the support, 'opening the square' (*kai ping fang* 開平方) is possible to extract the square root. The expression 'opening the square', appears in the descriptions of the Sections of Pieces [of Areas] procedures in Problems 22, 62 and 63, and in every solution with the Celestial Source procedure.

The most interesting part of the procedure is that it links the situation described in the statement and the setting up of the support with a diagram. The discourse inserted inside and around the diagram [1.9] gives explanations for the selected numbers. This discourse prompts an interpretation of the diagram as an assemblage of pieces representing the terms of the equation. As the following example shows, this assembly is conceptualised as a set of 'stacked areas'. The reader must visualise a second meaning of the diagram as components piled on top of one another, rather than as puzzle pieces fitted next to each other. This visualisation also constitutes a key point which separates the use of diagrams in the Chinese traditions with their use in the Euclidean traditions. In the present case, the apex is never marked or taken into account. The diagram is considered here as a pile of plain surfaces on which one has to operate. Liu Hui's diagram, as presented by Yang Hui and the diagrams of the *Development of Pieces [of Areas]*, seems different. In the first case, the pieces of areas are rearranged to be next to each other, while in the *Development of Pieces [of Areas]*, the pieces of areas are piled up, as if a type of third dimension had been added through the reading. Like what is seen in the *Gnomon of the Zhou*, there is no a unique figure which directs the reading of the discourse on the algorithm. The diagram evolves according to the construction of the algorithm, but the evolution used a new dimension. This feature is in opposition to the Euclidean version of diagram.

To understand the diagram of Problem 1, for example, one must begin with the data given in the statement which Li Ye represented in the first diagram. The distance a and the area A [Fig. 2.9] are known. With this data, pieces of square fields of side a can be identified such that on the given area A, four squares can be constructed with the given side-length a corresponding to a total area of $4a^2$, which is a constant [Fig. 2.10]. The purpose is to express the known area in term of what is unknown. Thus, Li Ye wrote '*Subtracting four pieces of the square of the reaching bu from the real area is to subtract four corners.*' When these squares are removed, the area that remains can be read as an expression in terms of the unknown. We thus have $A - 4a^2$ in [Fig. 2.10], and the cross-shaped area represents $4ax + x^2$ [Fig. 2.11]. However, this area does not correspond to the area given in the statement because the area of the circular pond still must be removed. This circular area equals when π is assumed equal to 3; that is to say that an area of $0.75x^2$ must be subtracted from the square in the centre: $x^2 - \frac{3}{4}x^2$. Thus, removing the circle leaves an area of $0.25x^2$ [Fig. 2.12]. Consequently, Li Ye wrote '*[taking] two fen and a half as the constant divisor, [it] is [found] that for each bu of the inside [part] full [of water], seven fen and a half are [taken] off, [and] outside there are two fen and a half.*' Thus, the diagram is read as $A - 4a^2 = 4ax + 0.25x^2$.

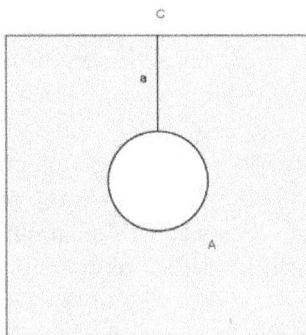

Fig. 2.9

2.2.3 *Interpretation*

Diagrams here do not contain universal or infinite space. The diagrams should be interpreted as a superposition of areas, showing that a known area is equivalent to the same area expressed according to the unknown. Thus, the diagram is the object of two readings; one translates the diagram into a constant, the other into an unknown. The diagram presents the figure described by the quadratic equation. The second reading of piled areas is the expression of the equality. The global area corresponds to the constant term, and drawing it is equivalent to writing that this value is equal to the squares representing the term in x and x^2. To trace a diagram

Fig. 2.10

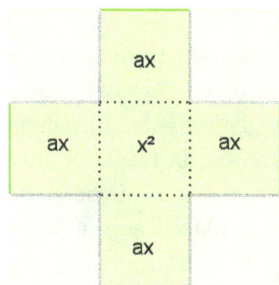

Fig. 2.11

is to express the equality, and to see the terms of the equation. Thus, Li Ye wrote in problems 23, 25 and 27 that the terms are *zi jian* 自見, literally 'self-visible' or better 'self-evident'. The diagram is in fact self-sufficient.

The diagrams are not mere illustrations; they are objects to be read. Naturally, they necessitate a special way of reading. The practitioner has to read into it, directly. To read the diagram and the 'meaning' accompanying it indicates how the data of the problem can be transformed. To draw the diagram is to trace the origins of the terms of the equation. The

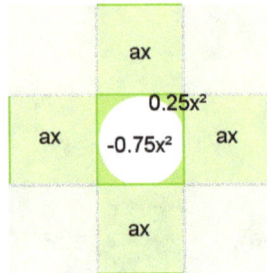

Fig. 2.12

computation of the different coefficients on the support is paralleled by the construction of the diagram. The practitioner can trace back the source of the constant divisor, which was given as a final result on the description of the counting support [1.8]. The role of the diagram is, for this part, heuristic because the procedure is used to discover the equation verified by the unknown. This research is made through the transformation of areas and the reduction of the area given in the statement to an assembly of pieces of areas that can be interpreted in terms of polynomials.

The diagram explains the origin of the area that is used for computation $(A - 4a^2)$ and which is placed as the dividend, as well as the source of the adjunct $(4ax)$ and the constant divisor $(0.25x^2)$. By legitimising the sources of the areas, it consequently confirms also the validity of the procedure. This reasoning also justifies the choice of the values $(A - 4a^2;$ $4ax; 0.25x^2)$ for the settings on the counting support. The same diagram verifies the validity of a given procedure, which relates to the computation of coefficients presented on the counting support. The diagram is at once an interpretation, a rewording of the data of the statement of the problem and a way to visualise the equation which is verified by the unknown. It provides verification of how the data of the statement are transformed into an equation, so its value is also demonstrative. The diagrams are used in the context of argumentation; so they do not appear only as mere illustrations but are rather a way to master the type of transformations of areas entering the reasoning. The diagram is the equation and the equation is the diagram.

Chapter 3

Transformation of Diagrams: The Example of Problem 21

Diagrams are not only objects to be published in mathematical treatises. The published materials are relics of set of practices. The diagrams are indeed mathematical object, but in the present case, they are also evidence for specific culture of discourse which relates to their philosophical background. In the *Development of Pieces [of Areas]*, diagrams were objects to be drawn, visualised as transitory. They are silent textual objects, part of speechless discourse.

Several comments written by Li Ye indicate that great care was given to the quality of the drawing. The following comment accompanies the diagram for Problem 45: '*But if [the drawing] is slightly inclined, then one cannot use it.*'[1]

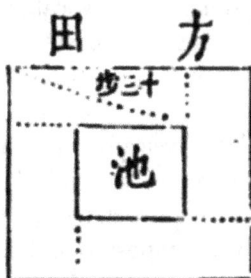

Fig. 3.1   Problem 45

The accuracy of drawings seems to have been an important issue. In evidence of this, Li Ye commented on the diagram of Problem 61 that '*the*

---

[1]若稍有偏側, 則不能用也.

two fen inside this diagram have to be drawn in an extremely thin shape.[2]

Fig. 3.2   Problem 61

Li Ye did not state why precision is required, nor did he explain why Li Ye felt it necessary to comment on it in these particular problems.[3] It is difficult to tell if these comments were intended as a description or as instructions for the reader. However, context indicates that the diagrams were not only intended to be examined but also to be drawn by the reader, and that their elements were intended to be objects of transformations.

As a consequence of these intentions, some areas are 'assembled', 'gathered' or 'decomposed' and some of the vocabulary used in the meanings also recalls physical manipulations: 'to paste' (*tie* 貼, Problem 34)[4] or 'to stack', 'to pile up' (*die* 疊, Problems 19; 20; 22; 24; 26; 52; 53; 58 and 59).[5] However, it is difficult to believe that sixty-four figures made of different components were meant to be physically manipulated, cut into pieces and piled together. Besides, it is *a fortiori* difficult to manipulate 'negative' empty areas. Thus, the reader faces a challenge in how to proceed in drawing the diagrams while also decomposing and assembling pieces of areas. An examination of Problem 21 may provide some clues.

---

[2]此圖內二分合畫作極細形狀.

[3]Also Problem 19: '*Now, one looks for the diagonal of the square; therefore, this diagram requires thin parts*' 今求方斜, 故其圖須細分之.

[4]Problem 34: '*Inside the eighth bu of the adjunct, one pastes eight squares of the bu of the reaching diagonal*' 八个從步內, 貼入八个斜至步羃.

[5]For example, Problem 20: '*On the bu of the adjunct, one stacks six hundred twenty-five squares of the diameter of the pond*' 於從步上, 疊用了六百二十五个池徑羃.

3.1 Translation

Problem 21.

[21.1] Let us suppose there are three pieces of square field. [Added] together, the area counts four thousand seven hundred seventy *bu*. It
is said only that the sides of the squares are mutually comparable[6]
and the sides of the three squares summed up together yields one
hundred eight *bu*.
One asks how much are the sides of the three squares each.
[...]

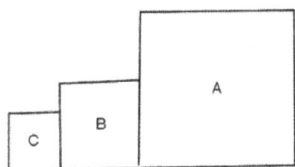

Fig. 3.3   *A*: big square; *B*: middle square; *C*: small square.

[...]
[21.6] One looks for this according to the [procedure of the] Section of
Pieces [of Areas]. Place the quantity of the sum [of the area of
the three fields]. What results once divided by three is the side of
the middle square. Self [multiply] this to make the square. Triple
this further and subtract this from the area to make the dividend.
There is no adjunct. The constant divisor is two *bu*.
[21.7] The meaning: from the *bu* of the area, one subtracts three squares
of the middle square. Outside there are two squares. Therefore, it
yields two *bu*, the constant divisor.

3.2 Description and Interpretation

[21.1] states: let a, b and c be the respective sides of the squares A, B, C.
Let their sum be equal to 108 *bu*, the sum of $A + B + C = 4770$ *bu*, and
$c - b = b - a = x$. It is required to find x, the difference between the sides
of the squares.

As in Problem 1, we can synthesise data as [Table 3.1].

---

[6]方方相較: the difference between the side of the small square and the side of middle
square equals the difference between the side of the big square and the side of the middle
square.

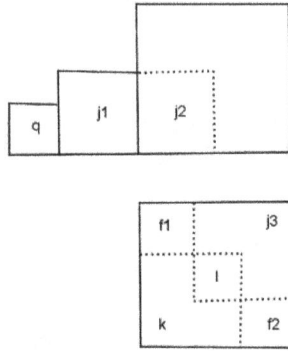

Fig. 3.4   j1-3: subtract; q: to go to; l: to come to; k: empty; f1, f2: square.

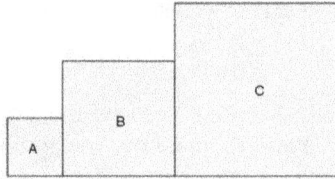

Fig. 3.5

Table 3.1:

$A + B + C$–$3b^2$	dividend
$\emptyset$	adjunct
2	constant divisor

From [21.6], the equation may be stated as: $A + B + C - 3b^2 = 2x^2$.

The first step is to interpret the data given: an area equal to $A + B + C$ and a distance equal to $a + b + c$. Since $c - b = b - a$, one infers that $\frac{a+b+c}{3} = b$. Thus, the first step of the procedure is to express each of the areas according to b. (see [Fig. 3.6]), that is, $B = b^2$, $A = b^2 - (2bx - x^2)$.

To express A, one removes from b^2 a gnomon made of two rectangles stacked on one square. These two rectangles reify what is unknown: their length is b, and their width is x. Or equivalently, $b^2 = A + 2bx - x^2$, $C = b^2 + 2bx + x^2$. To make C, one adds to b^2 a gnomon made of two rectangles, with length b and width x. To this another square of side x is added at the corner to complete the area. Therefore, each of the squares has been expressed according to the constant and the unknown identified in the statement.

The second step is to remove three squares of side b in order to construct the constant term, that is, $A + B + C - 3b^2$. Li Ye wrote in [21.7] '*from the bu of the area, one subtracts three squares of the middle square.*' In Li Ye's diagram, two of the squares are marked by the character 'subtract' (*jian* 減) and are removed first, one from B and one from C (see [Fig. 3.7]). The problem is now to remove the third square of side b. To remove this third square is equivalent to removing $A + 2bx - x^2$. If one re-assembles the elements and recomposes the diagram, [Fig. 3.8] results. That is, a square of side c, from which b^2 has been removed once (this area is marked 'void', *kong* 空, by Li Ye), on which is 'stacked' a square of side a in the middle, with a gnomon composed of two rectangles of length b and width x; from the latter, a square of side x is removed (See construction of A according to b). Once this third square has been removed, there remain two squares of side x at two of the corners. Thus Li Ye wrote in [21.7]: '*outside there are two squares*'. See [Fig. 3.9]. The diagrams has thus represented $A + B + C = b^2 + (b^2 - 2bx + x^2) + (b^2 + 2bx + x^2)$ and transformed this into $A + B + C - 3b^2 = 2x^2$.

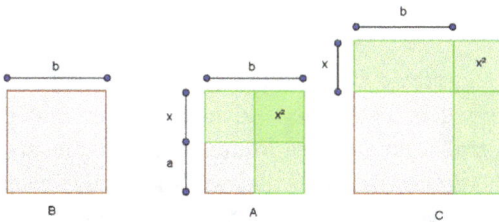

Fig. 3.6

Li Ye chose to represent the two steps of the procedure with two diagrams, whereas the discourse preserves a total of three diagrams for this

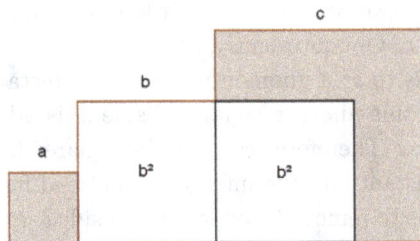

Fig. 3.7

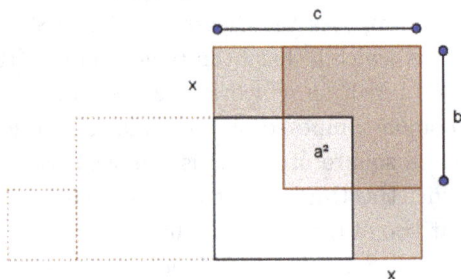

Fig. 3.8

problem:

(1) The diagram illustrates the statement.

(2) A diagram in which the shape is identical to the first one and which operates as a transitional phase from the data of the statement to the first step of the procedure. This diagram shows the two areas which are to be removed, as well as that the small square, marked by the character 'to go' (*qu* 去) which serves as the object of the next operation.

(3) A diagram shows the destination of the small square, marked by the character 'to come' (*lai* 來). The areas which were subtracted are represented by dotted lines. The third diagram is the result of imaginary manipulations: it justifies the origin of its terms by helping the reader to visualise the equation.

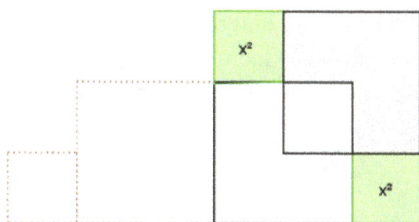

Fig. 3.9

Thus, three diagrams can present an algorithm for setting up an equation, if the diagram illustrating the statement is included. Considering this evidence, the reader was supposed to draw the diagram while imagining the movement of transformations which lead from the first to the final diagram. Here, 'mental' or 'imaginary transformation' should be defined as a way of visualising and following transformations on a drawn geometrical figure. This conclusion shows that the textual part of the Section of Pieces [of Areas] is the tip of iceberg for non-discursive practices. Reading the text is a small part of the work of the practitioner; the main part of the work consists of drawing and visualising diagrams and manipulating counting rods.

Here, the work by Li Ye on diagrams shows both continuity and a rupture with tradition. Chemla ([Chemla (2001)], [Chemla (2010)]) showed that in third century China, there was a practice wherein diagrams were cut and materially reassembled, as shown in the commentaries to the *Gnomon of the Zhou* and the *Nine Chapters*. In the *Development of Pieces [of Areas]*, diagrams were inserted as visual artefact inside a text, no longer intended to be physically manipulated. They are closer to what [Volkov (2007), 441] describes concerning interpretation of geometrical operations. Diagrams are still objects to aid the visualisation of imaginary transformations on a drawn support. Yet, they are no longer physical objects to be displayed outside of the text. Movement of piled areas is the key to understanding Chinese diagrams. Visualisation is not just a matter of looking at something. Visualisation embodied element intentionality. There is also a goal: the perception of movement is used to construct an invisible object — invisible because it is nothing but movement. Visualisation requires imagination. Here, diagrams are not representations of ideal objects in which objects have to be imagined. The practitioner undertakes the act of the construction literally within the diagram. This practice involves the cultivation of a mode of attention. Geometrical figures as a dynamic element creates a technique which consists of directing attention toward the object.[7]

---

[7]This point will be developed in part III.

Conclusion of Part I

From the examples of problems 1 and 21, it is possible to order the steps in the practice of using diagrams and texts. First, the data of the problem were read; then, the practitioner manipulated the rods and drew a diagram. The last step was to transcribe the terms of the equation into sentences such as those of [1.8] or [21.8], where the results of computation were given *in fine*. At this step, the counting board was set up to directly extract the square root. Diagrams were drawn by readers before the text was written. The diagrams we see now in the edition of *Development of Pieces [of Areas]* were added by Li Ye at the very end of the process, as an editorial practice.

The Section of Pieces [of Areas] procedure functionally rewords the conditions of the statement in terms of the configuration which casts light on the link between the known and unknown quantities of the problem. It then proceeds to decompose and regroup the elements of the diagram in order to visualise the fundamental identity proposed as the resolution — an identity between a numerically known area and an assembly of areas expressed with the unknown. The Section of Pieces [of Areas] establishes the link between a diagram and an arithmetical resolution. The solution of a problem is articulated by the relation between a diagram showing an identity and a tabular setting, for which the modality of use was well known at the time Li Ye was writing. The diagram indicates the equality and the articulation of the terms of the equation, while the numbers appearing on the counting support indicate the quantities of the coefficients. This correspondence functions as an equation. The diagram is not a symbolic representation of the equation but the equation itself. To draw a diagram with accuracy is to construct an equation. To visualise the transformations of areas is to justify the construction of the terms of the equation. There is only one diagram illustrating the Section of Pieces [of Areas] procedure but this unique diagram ought to be read in terms of a pattern of transformations.

The examples given in Li Ye's *Development of Pieces [of Areas]* show that diagrams are not mere illustrations of the text and that the study of practices must be taken into account in the process of interpreting the mathematical objects. In this example, the diagram is the main tool for the construction of an equation. To draw and to visualise the articulation between pieces of areas represented in the diagram is the way to shape the equation. The discourse provides some hints to make this construction possible, but this text is not sufficient to understand the construction of the equation, whereas the diagram is. The diagram can function independently because the image supplies sufficient information. The diagram is part of the argumentation; it is involved in the process of justification of the coefficients of the equation. The same diagram can have several functions: heuristic, edificatory and demonstrative. It shows the correctness of an algorithm and it is transformed along with the algorithm's steps. From an ontological point of view, the diagram is not a schematic representation of an ideal perfect mathematical object. The diagram is not an imperfect representation depending on human perception opposed to a perfect mathematical object conceived by the reason. In the cases taken from the *Development of Pieces [of Areas]*, it is no less than the mathematical object itself and its transformations. It has a literal value. If an equation is a statement containing data, unknown and relation of equality, then the diagram is the equation. And the equation takes the shape of a diagram.

The practitioner must step beyond phenomena or appearances. They have to surpass words and their codes titled 'meaning'. The discourse 'speaks' about something beyond what they address, because, paradoxically, that is the best way to communicate what cannot be verbalised. Visualisation here is the best access to mathematical abstraction, as words are still connected to contingency. Mathematical proofs are expressions of necessity, while signs and meanings are linked arbitrarily. Here, language works with detachment: the finger is not the moon. As well, the diagram is not exactly the proof. It is the way to the proof, and a sign of it. To prove the correctness of an algorithm, to justify the construction of equation is to construct something through imagined actions. Therefore, it is necessary that the practitioner recapitulate the process and establish meanings independently. They have to see for themselves.

This use of language is the fruit of a reflection on language. Here, diagrams are the synecdoche by which a sign is used for a whole meaning. Saying that a diagram is a synecdoche (or metonym in [Netz (1999)]) is not a metaphor! It means that, as part of text, they have a semiotic status. Inasmuch language is an access to the world, and may be, nothing less than the world itself, these diagrams are keys to accessing reality. The ramifications for ontological questions must be left aside for the moment.

The purpose of this discourse by diagram and characters is to communicate that discourse in its totality is just figures. This rhetoric of provisional diagram seeks resolution in the creation of a discursive figure. The proof can be considered as a discourse on figures. The diagram should be understood without interpretation. It ought to be understood directly, without superimposed interpretations or textual explanations. Its truth relies builds upon what the practitioner personally understands. Because diagram embodies movement, understanding of it is a dynamic act. Diagrams are imbued with efficiency of a direct access to truth which works in a visual mode instead of an auditory mode. The work of Li Ye places in evidence an attempt of rationalisation and organisation of knowledge. Diagrams are a silent link between thinking and words. Words lead to pictures, which in turn must lead to the thoughts contained in them, and through the diagrams, the thoughts have a tangible form. It is not enough to tell things. Things must be shown. Li Ye deliberately uses metaphors, even while he expects the practitioner to think beyond metaphors.

Problem 1 and Problem 21

第一問[8]

[1.1] 今有方田一段, 內有圓池水占, 之外計地一十三畝七分半. 竝不記內
圓外方. 只云從外田楞至內池楞, 四邊各二十步.
問內圓外方各多少?
荅曰: 外田方六十步, 內池徑二十步.

[1.2] 法曰: 立天元一為內池徑. 加倍至步得 為田方面.

[1.3] 以自增乘得 為方積, 於頭.

[1.4] 再立天元一為內池徑. 以自之, 又三因, 四而一得 為池積.

[1.5] 以減頭位, 得 為一段虛積, 寄左.

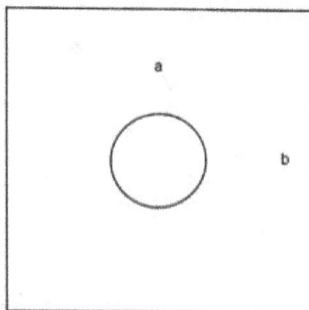

Fig. 3.10   a: 至水二十步 b: 方田六十步.

[1.6] 然後列真積. 以畝法. 通之, 得三千三百步. 與左相消

---

[8]The commentaries were removed from the discourse. To read their translations see
[Pollet and Ying (2017)]. The present version of the Chinese text is based on Li Riu
(LR) edition which was carefully compared to two of the edition of the Complete Library,
namely the Wenjing (WJG) and Wenyan (WYG) edition. See [Pollet (2014)] for the
detail of the comparison.

[1.7] 得 [⚏] ⁹ 開平方, 得二十步, 為圓池徑也. 倍至步, 加池徑, 即外方面
也.

[1.8] 依 ¹⁰ 條段求之. 真積內減四段至步冪為實. 四之至步為從. 二分半
常法.

[1.9] 義曰: 真積內減四段至步冪者, 是減去四隅也. 以二分半為常法者,
是於一步之內占, 却七分半, 外有二分半也.

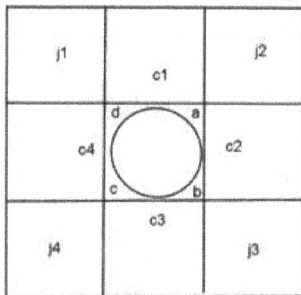

Fig. 3.11 j1-4: 減; c1-4: 從; abcd: 二分五厘.

Translation:

Problem 1.

[1.1] Let us suppose there is one piece of square field, inside which
there is a circular pond. Outside the [area] occupied by water, one
counts thirteen *mu* seven *fen* and a half of land. Moreover, there
is no record of the [dimensions] of the inner circle and the outer
square. It is said only that [the distance] from the edge of the
outer field reaching the edge of the inside pond on [all] four sides
is twenty *bu*.

One asks how much are [the diameter of] the inner circle and [the
side of] the outer square. The answer: the side of the outer field is
sixty *bu*. The diameter of the inside pond is twenty *bu*.

[1.2] The method: set up one Celestial Source as the diameter of the
inside pond. Adding twice the reaching *bu* yields $\begin{matrix} 4\,0\ tai \\ 1 \end{matrix}$ as the

---

⁹The first number is 1700 in WJG and WYG instead of 700 in LR.
¹⁰以 instead of 依 in WYG and WJG.

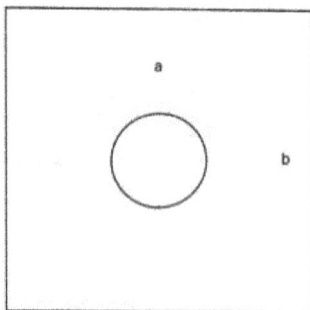

Fig. 3.12   *a*: The *bu* reaching the water are twenty; *b*: The side of the field is sixty *bu*.

side of the field.[11]

[1.3] Augmenting this by self-multiplying yields
$$\begin{matrix} 1\,6\,0\,0 & tai \\ 8\,0 \\ 1 \end{matrix}$$
as the area
of the square, which is sent on the top.

[1.4] Set up one Celestial Source as the diameter of the inside pond. This times itself and multiplied further by three then divided by
four yields
$$\begin{matrix} 0 & tai \\ 0 \\ 0\,.\,7\,5 \end{matrix}$$
as the area of the pond.

[1.5] Subtracting this from the top position yields
$$\begin{matrix} 1\,6\,0\,0 & tai \\ 8\,0 \\ 0\,.\,2\,5 \end{matrix}$$
as
a piece of the empty area, which is sent to the left.

[1.6] Next, place the real area. With the divisor of mu, making this communicate yields three thousand and three hundred *bu*.

[1.7] With what is on the left, eliminating from one another yields
$$\begin{matrix} 1\,7\,0\,0 & tai \\ -\,8\,0 \\ -\,0\,.\,2\,5 \end{matrix}\,.$$
Opening the square yields twenty *bu* as diameter of the circular pond. Adding twice the reaching *bu* to the diameter of the pond gives the side of the outer square.

[1.8] Look for this according to the Section of Pieces [of Areas]. From the real area (真積), four pieces of the square of the reaching *bu* (四段至步冪) are subtracted to make the dividend (實). Four times the reaching *bu* (四之至步) makes the adjunct (從). Two *fen* and a

---

[11]田方面, *Tian fang mian*. The use of the character '*mian*' at this place is very rare in this treatise; the character '*fang*' is usually sufficient to mean the side.

half is the constant divisor (常法).

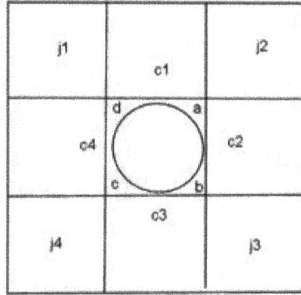

Fig. 3.13   j1–4: subtract; c1–4: adjunct; abcd: two *fen* five *li*.

[1.9]The meaning: From the real area, to subtract four pieces of the square of the reaching *bu* is to subtract four corners. [Taking] two *fen* and a half as the constant divisor, is that for each *bu* of the inside [part] full [of water], seven *fen* and a half are [taken] off, outside there are two *fen* and a half.

第二十一問

[21.1] 今有方田三段, 共計積四千七百七十步. 只云方方相較等. 三方面
共併得一百八步. 問三方各多少.
荅曰: 大方面五十七步. 中方面三十六步. 小方面一十五步.

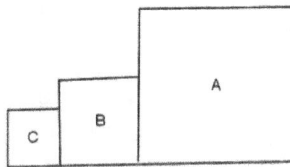

Fig. 3.14   a: 大方; b: 中方; c: 小方.

[21.2] 法曰: 立天元一為方差. 以減中方面置併數三而一即得中方面得

為小方面也.

[21.3] 以自之得 ⬚ 為小方積, 於頭. 再立天元方差.

[21.4] 加入中方面得 ⬚ [12] 為大方面.

[21.5] 以自之得 ⬚ 為大方積, 於次位.

[21.6] 又列中方面 ⬚ , 自之得下 ⬚ 為中方積, 於下位三位. 相併得

⬚ 為一段如積數, 寄左. 然後列真積四千七百七十步.

[21.7] 與 左 相 消 得 ⬚ .   開平方得二十一步, 即是方差也.
置方差數. 加中方, 即大方面. 減中方, 即小方面也.[13]

[21.8] 依條段求之. 列併數. 以三約之, 所得即中方面也. 以自之為冪. 又
三之. 以減積為實. 無從. 二步常法.

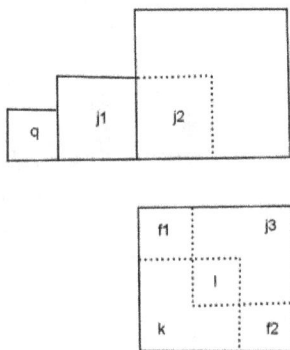

Fig. 3.15   j1–3: 減; q: 去; l: 來; f1–2: 方; k: 空.

[21.9] 義曰: 積步內減三个中方冪. 外有兩个方. 故得二步常法.

[21.10] 舊術: 又折半. 止得一个方也.

Translation:

---

[12]太 not in WYG.

[13]The sentence underlined is presented as a commentary in WYG and WJG.

Problem 21

[21.1] Let us suppose there are three pieces of squares fields. [Added] together the area counts four thousand seven hundred seventy *bu*. It is said only that the sides of the squares are *mutually compara-ble*[14] and the sides of the three squares *summed together* yields one hundred eight *bu*.

One asks how much are the sides of the three squares each.

The answer: the side of the big square is fifty seven *bu*, the side of the middle square is thirty six *bu* and the side of the small square is fifteen *bu*.

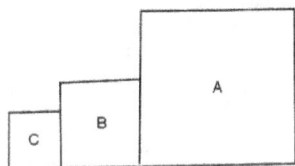

Fig. 3.16   A: big square; B: middle square; C: small square.

[21.2] The method: set up one Celestial Source as the difference between the sides.[15] Subtracting it from the side of the middle square yields $\begin{array}{l} 3\ 6\ tai \\ -\ 1 \end{array}$ as the side of the small square.

[21.3] This times itself yields $\begin{array}{l} 1\ 2\ 9\ 6\ tai \\ -\ 7\ 2 \\ 1 \end{array}$ as the area of the small square, which is sent to the top.

[21.4] Set up again the Celestial Source, the difference between the sides. Adding the side of the middle square yields $\begin{array}{l} 3\ 6\ tai \\ 1 \end{array}$ as the side of the big square.

[21.5] This times itself yields $\begin{array}{l} 1\ 2\ 9\ 6\ tai \\ 7\ 2 \\ 1 \end{array}$ as the area of the big square, which is placed on the next position.[16]

---

[14]方方相較: the difference between the side of the small square and the side of middle square equals the difference of the side of the big square and the side of the middle square.

[15]方差, *fang cha.*

[16]*Yu ci wei.* This problem requires to place 3 polynomials on the table, the names of the position are thus different.

[21.6] Place further the side of the middle square, 36 *tai*. This times itself yields 1296 *tai* as the area of the middle square, which is sent at the bottom position.[17]

Mutually adding the three positions yields

$$\begin{array}{c} 3\ 8\ 8\ 8 \\ 0^{\,18} \\ 2 \end{array}$$ as one piece

of the quantity of the equal area, which is sent to the left.

Afterwards, place the real area, four thousand seven hundred seventy *bu*.

[21.7] With what is on the left, eliminating from one another yields

$$\begin{array}{c} -\ 8\ 8\ 2 \\ 0 \\ 2 \end{array}$$

Opening the square yields twenty one *bu*; that is the difference between the sides. Put down the quantity of the difference between the sides and add the side of the middle square; it gives the side of the big square. Subtract the side of the middle square; it gives the side of the small square.[19]

[21.8] One looks for this according to the section of pieces [of areas]. Place the quantity of the sum. What results once divided by three is the side of the middle square. Self [multiply] this to make the square. [Multiply] this further by three and subtract this from the area to make the dividend. There is no adjunct. The constant divisor is two *bu*.

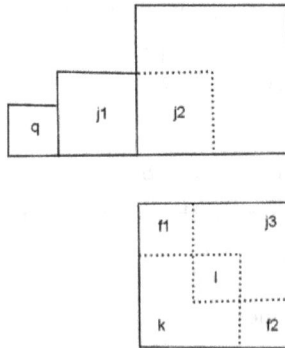

Fig. 3.17   j1–3: subtract; q: to go to; l: to come to; k: empty; f1, f2: square.

---

[17] *Yu xia wei.*

[18] The character *tai* is not written here.

[19] The two last sentences are presented like a commentary in WJG and WYG *siku quanshu.*

[21.9] The meaning: from the *bu* of the area, one subtracts three squares of the middle square. Outside there are two squares. Therefore, it yields two *bu*, the constant divisor.

[21.10] The old procedure: One only reduces further to the half. It yields one square.

Problem 21, description.

[21.1] Let a, b and c be the respective sides of the squares A, B, C. Their sum equal to 108 *bu*; let the sum of $A + B + C = 4770$ *bu*; and $c - b = b - a = x$.

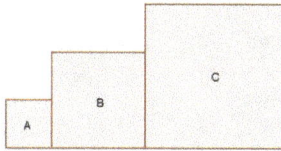

Fig. 3.18

The procedure of the Celestial Source:

[21.2] $\frac{a+b+c}{3} = b, 108/3 = 36$.

[21.3] $A = (b - x)^2 = b^2 - 2bx + x^2 = (36 - x)^2 = 1296 - 72x + x^2$.

[21.4] $B = b^2$.

[21.5] $C = (b + x)^2 = b^2 + 2bx + x^2 = 1296 + 72x + x^2$.

[21.6] $A + B + C = 3b^2 + 2x^2 = 3888 + 2x^2 = 4770$.

[21.7] The equation: $3b^2 - (A + B + C) + 2x^2 = -882 + 2x^2 = 0$.

The old procedure: [21.10]

$$\frac{(A + B + C) - 3b^2}{2} = x^2.$$

Part II: Operations on Pieces of Areas

Movement is a key to understanding the diagrams of the *Development of Pieces [of Areas]*. Several different possible transformations can be observed in the diagrams of the *Development of Pieces [of Areas]*. Not only areas are moved virtually and cut into pieces but they are also filled and emptied. The relation between plain and empty areas lies at the core negative and positive coefficients. Indeed, the question is how to represent a negative coefficient by mean of areas. Through a comparison of Li Ye and Yang Hui's works, it is possible to reconstruct some steps in the development of the elaboration of quadratic equation. It is also possible to distinguish several layers in the composition of the text transmitted by Li Ye. This section draws upon Problems 18 and 5 of the *Development of Pieces [of Areas]* and Problem 49 of Yang Hui's Methods of Computation. Problems 18 and 5 offer cases from which the Old Procedure can be reconstructed and enable the interpretation of the Section of Pieces [of Areas] as containing several procedures depending on the different reading and cutting of the diagrams. In light of Problem 49 of *Yang Hui's Method of Computation*, a clear evolution of concept of negative coefficient emerges. To understand the paths of these transmission and interpretation, it is first necessary to visualize the different operations of transformations as performed on diagram (illustrated by Problem 2).

Chapter 4

The Six Transformations: The Example of Problem 2

The reader of the *Development of Pieces [of Areas]* is supposed to visualise several transformations of the diagrams in order to set up equations. The number of transformations is not infinite, and these transformations are carefully chosen. It is possible to inventory each type of transformation and describe their mathematical meaning. In fact, the first chapter of the *Development of Pieces [of Areas]* uses only six types of transformations for the Section of Pieces [of Areas]. These transformations work like fundamental operations of arithmetic. They are the six models of bricks with which it is possible to construct every geometrical algorithm. In the other two chapters of the treatise, the operations increase in sophistication and operate on objects that are more complex but essentially remain the same. These operations appear after the reader has gained experience in reproducing problems at the point when the pattern of vocabulary and drawing become repetitive. This chapter establishes elements necessary to Chapter 7 wherein the combination of operations becomes meaningful.

Here are the six basic geometrical operations used to transform diagrams in the *Development of Pieces [of Areas]*:

A: removing corner(s) (*jian* 減).
B: piling and unstacking areas (*tie* 貼, *die* 疊)
C: compensating areas (*bu* 補)
D: expanding areas (*zhan* 展)
E: multiplying by parts (*fen* 分)
F: moving an area, cancelling areas (*qu⋯lai* 去⋯來, *luo* 漏)

The first operation, named A — 'to remove corners' — is the most basic and is involved in each of the problems. It consists in cutting some areas from the diagram. Problem 1 discussed earlier clearly illustrates a case where four square 'corners' are removed from a figure. Usually, several similar areas are removed from the figure: 4 squares, 6 rectangles, etc.; they simply disappear from the visual representation. This operation evolves after its introduction in chapter 1. In chapter 2, it becomes clear that there are several ways to remove areas, depending of the sophistication of the object, which needs to be removed. Operation A deals with simple squares, rectangles, or with gnomons made of four rectangles and a square. Depending on the algorithm, either the whole figure of the gnomon is removed like the corners in Problem 1 (Problem 32), or the gnomon can be removed by steps (For example, two rectangles may be removed first, then two other rectangles, and finally the square) (Problems 34, 35) or the removal may be conceptualised without being drawn (Problem 36). The complexity of an operation is denoted by double letters, i.e. AA [see Table A.3 in Appendix A.1] and the type of the objects on which it operates by the notation 'type 1, 2 or 3'. Thus, the fundamental operation A of chapter 1 can manifest as 'AA: To subtract a gnomon (type 1)' which denotes that the whole gnomon is directly removed like a square corner in A. The designation 'type 2' clarifies that the gnomon is removed in two steps/parts (the rectangles are removed first, then the square inside); while type 3 means that the removal is conceptually implied in the construction before drawing, as occurs in chapter 2 and 3.

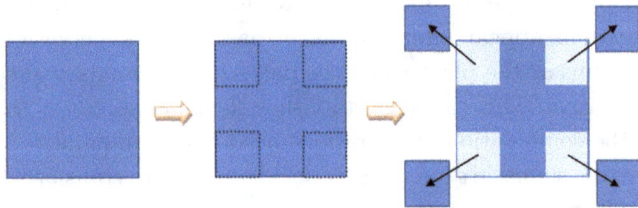

Fig. 4.1   Transformation A

The second operation, B — 'to pile and unstack' — is illustrated by Problem 2 at the end of this chapter. It consists of 'opening' areas like a book. When two areas are piled together, it is possible to display the two areas next to one another, like an opened book. Since there is twice the same area, one of the areas is considered an unnecessary duplicate which must be removed by cutting. This operation also evolves in complexity together with the objects on which the operations acts. If one area is first removed, followed by the remaining area being cut or partially cut (Problem24), this operation is denoted as 'BB: to unstack twice'. As with operation A, several different objects may be governed: type 1: circle (Problem 46), type 2: empty external square (Problem 53), type 3: full internal square (Problem 29) and type 4: protruding square (Problem 30). Problem 18 of chapter 5 offers another illustration of this operation.

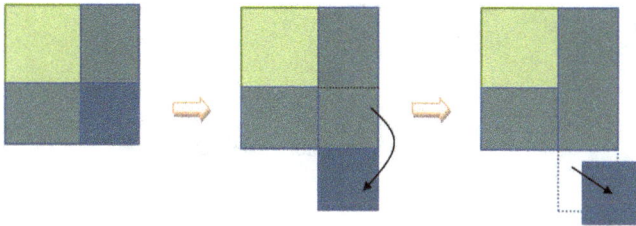

Fig. 4.2   Transformation B

The third operation, C, means 'to compensate'. This operation is illustrated in the next chapter by Problems 18 and 14. It consists of adding back the missing part of an area when too much area has been removed by operations A or B. This operation always deals with simple objects in all chapters.

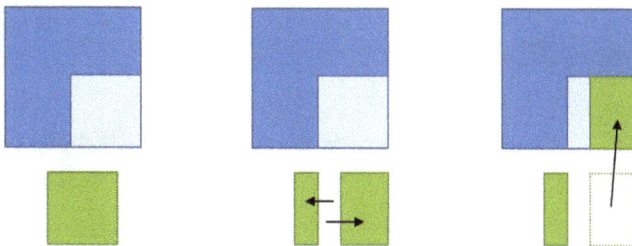

Fig. 4.3   Transformation C

The fourth operation, D — 'to expand', transforms the diagonal of a square into the side of a bigger square through multiplying by $\sqrt{2}$, where $\sqrt{2} = 1.4$. Problem 3 illustrates this operation in Chapter 7. This operation evolved in the course of chapters treating different objects: type 1 consists of expanding the outer figure without expanding the inner one (Problem

39); type 2 consists of expanding the interior figure without expanding the outer one (Problem 50) and type 3 approximates with 1.2 instead of 1.4 (Problems 61 and 62).

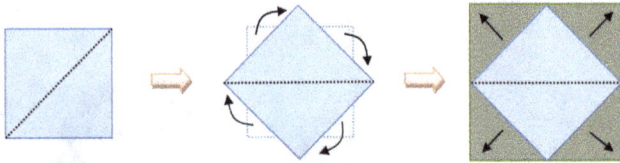

Fig. 4.4   Transformation D

The operation E, 'to multiply by the parts', transforms a circle into 3 squares, or multiplies each coefficient by what Li Ye calls a 'denominator'. By assuming that $\pi = 3$, it is possible to convert four circular areas into three squares with sides equal to their diameters. This conversion makes the geometrical transformations easier. Sometimes, instead of multiplying by 3, all coefficients are multiplied by the same factor to fit another area. Problem 5 given in Chapter 6 illustrates this operation.

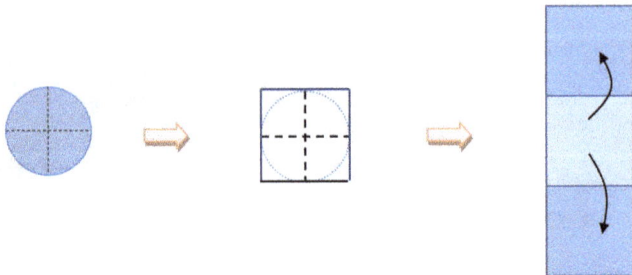

Fig. 4.5   Transformation E

The last operation F, 'to move an area' or 'to cancel areas', comprises shifting an area from one place to another. Then, the two areas are cancelled from one another. This operation was previously illustrated by Problem 21 in Chapter 3. This operation reached a higher degree of sophistication in Problems 31, 40, 41, 42 and 54 with the result that the operation was renamed 'to diffuse' (FF in Table A.3 in Appendix A.1).[1] In these four problems, it is not possible to draw the cancellation as in Problem 21. The reader must imagine that one area is cancelled from another area, as if one of the areas was spread inside the other. The cancellation is performed arithmetically on the counting surface. Li Ye represents this operation in Problems 40 to 42 by drawing half a circle in and half out of the square [See Fig. 4.7. Problem 41] [See also Fig. 7.12. Problem 42 in Chapter 7]. This circle is not the representation of an area which must be removed or piled from/to another one. It is in fact 'diffused' inside another area, like a piece of sugar melting in water.

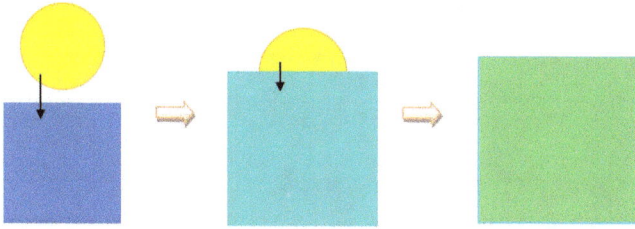

Fig. 4.6   Transformation F

Fig. 4.7   Problem 41

---

[1]The literal and correct translation of *luo* 漏, is 'to leak'. Because the problems concern the areas of ponds full of water, the literal translation would be misleading to the understanding of the problem. Thus, *luo* is translated as 'to diffuse'.

Using these six transformations, the practitioner 'plays' with different type of areas and imagines how they are displayed and transformed in order to create an equation. However, difficulties arise when the problem involves the use of a negative coefficient. How should a negative area be described? Is there any concept of a negative area? Is it possible to draw an empty area?

Chapter 5

Geometric Representation of Negative Coefficient: The Example of Problem 18

How may the equations in the Section of Pieces [of Areas] procedure be understood? How may an equation be rendered into modern notation when the statement of equality, the designation of the unknown variable, and the sign of numbers are unwritten? The *Development of Pieces [of Areas]* does not explain how negative and positive quantities are differentiated. There are no traces of a symbolic notation. The characters *fu* 負, 'negative' and *zheng* 正, 'positive', usually name negative or positive quantities but they do not appear in the text concerning the Section of Pieces [of Areas].[1] These characters *fu* and *zheng* appear only in the commentary on Problem 14 by Li Ye, and not in the procedure itself: '[Whenever there is] a negative adjunct (*fu cong*) with a positive corner (*zheng yu*) or a positive adjunct (*zheng cong*) with a negative corner (*fu yu*), the dividend (*shi*) remains always the same. That is why the edge and the adjunct are used to differentiate'.[2] Conversely, the term 'empty' (*xu* 虛) and 'augmented' (*yi* 益) are associated with the expressions transcribed as negative quantities in modern mathematical terms. In the Celestial Source procedure, the character *xu* appears twice in Problems 1 and 2.[3] In these cases, the character names an area expressed according to a polynomial composed of

---

[1] In the Celestial Source procedure, a diagonal stroke marks the last digit of a term written in rod numeral notation, but this sign was added by the eighteenth century editor, Li Rui. (Pollet 2014).

[2] '從負, 隅正, 或從正, 隅負, 其實皆同. 故因此廉從以別之'. An interpretation of this commentary appears later.

[3] An example from Problem1: 'Subtracting this from what is on the position yields [⋯] as one piece of an **empty** area which is sent on the left'. 以減頭位得 為一段虛積, 寄 左. This character *xu* appears in Problem 2 for the same sentence. These are the only two occurrences of such a use of this term. In other problems, the expression is 'equal area' (*ru ji* 如).

two terms of different powers. This expression of the area will be cancelled by its expression by a constant term to make the equation. How can a character used to name a negative coefficient in one procedure later be used to name entire polynomials in another procedure of the same treatise? Are these characters synonyms for *fu*, 'negative'? Their semantic range must be considered as well as their transcription into modern notation. This transcription depends on knowledge of the sign of a given term. On this point, it is interesting to look at how historians recorded negative terms.

Lam Lay-Yong [Lam (1984), 260] noted that for an equation of the form $ax^2 + bx = c$, 'if any of the terms a or b are negative, then the term *xu* or *yi* is prefixed'. Kong Guoping [Kong (1999), 98–100; 178; 183] also interpreted *yi* and *xu* as synonyms of *fu*, 'negative'. Martzloff [Martzloff (1987), 222] translated the character *yi* as 'negative' in his explanation of Qin Jiushao's extraction of square roots.[4] Indeed, by this reading, the first sentence of the description of the Section of Pieces [of Areas] procedure echoes modern mathematical terms in establishing an equation between a constant term and a polynomial composed of two terms of x and x^2. In this way, the terms *yi* and *xu* would coincide with the idea of negative coefficients.[5] How one interprets the characters and the mathematical concepts with which they are correlated directs the transcription of the equation into modern mathematical terms.

If the reading of *yi* and *xu* as 'negative' is adopted, these interpretations ought to conform to the different transcriptions found in secondary literature. Historians do not always transcribe the same equation in the same way: the signs and the equation can be interpreted differently. As the equation is never directly stated in the *Development of Pieces [of Areas]*, the same expression can be transcribed as $ax^2 + bx - c = 0$; $ax^2 + bx = c$; or $bx = c - ax^2$. In the first case, the constant term is negative. In the second case, there are no negative quantities, and in the third case, the second-order indeterminate term is negative. An examination of how historians have read the equations of the Sections of Pieces [of Areas] will introduce an examination of how Li Ye writes about what have later been identified as negative terms and equality.

Mei Rongzhao [Mei (1966), 140] did not discuss the meaning of *xu*

---

[4] The character *xu* is not used in Qin Jiushao's works.

[5] Here, follow the occurrences of *yi* and *xu* in the *Development of Pieces [of Areas]*. *Xu yu* 虛隅: Problems 3, 5, 11a, 14, 46 and 51; *yi yu*, 益隅: Problems 3, 10, 26, 29 and 55; *xu chang fa*, 虛常法: Problems 2, 22, 24, 29, 30, 41, 42, 54, 57 and 61; *xu cong*, 虛從: Problems 7 and 17; *yi cong*, 益從: Problems 10 and 14.

or how negative quantities are recorded, because he presented only examples that contain positive quantities. The only reference to negative terms appears in his descriptions of equations in the form $ax^2 + bx = c$, where $c > 0$, $b \geq 0$ and $a > 0$ or $a < 0$. Xu Yibao [Xu (1990), 69] also presented the equations as a comparison between a constant term and a polynomial. For example, in the portion of Problem 5, which literally reads 'From forty-eight pieces of the area of the field, one subtracts three pieces of the square of the *bu that does not attain* to make the dividend. Six times the difference makes adjunct. One is the empty corner'[6], is presented as $-x^2 + 6 \times 168x = 48 \times 13.2 \times 240 - 3 \times 168^2$. Xu Yibao did not mention the presence of *xu* or *yi*. Similarly, Kong Guoping [Kong (1988)] transcribed the equations as equivalences between a constant and a, the indeterminate terms. The equation from Problem 8, which reads 'from the square of the *bu* of the sum, one subtracts sixteen times the real area to make the dividend. Six times the *bu* of the sum makes the adjunct. Three *bu* is the constant divisor' is represented as $3x^2 + 6 \times 300x = 300^2 - 16 \times 3300$.[7] Like Mei Rongzhao [Mei (1966)], Kong Guoping rendered the equations as expressions of $ax^2 + bx = c$, where $c > 0, b \geq 0$ and $a \neq 0$.

Guo Xihan [Guo (1996)] transcribed the equations of *Yang Hui's Methods of Computation* in the same way as Xu Yibao and Kong Guoping. For instance, his 'Problem 9' (Problem 46 in Lam Lay-yong) is transcribed as $60x - x^2 = 846$, where $-x^2$ is named 'augmented corner' (*yi yu* 益隅) by Yang Hui. Guo Xihan commented that the '*yi yu* represents the second coefficient as equal to -1'.[8] Neither the meaning nor the geometrical representation of *yi* or *fu* is discussed. Lam Lay-Yong [Lam (1977)] cautiously translated *Yang Hui's Methods of Computation* by separating the terms *yi* and *fu*. In the translation, Lam Lay-Yong adopted the Wade–Giles transliterations *i yü* or *fu yü* (*yi yu* and *fu yu* in *pinyin*) as loanwords and noted 'Yang Hui's ingenious handling of negative terms' [Lam (1977), 259]. In the commentary and discussion, Lam Lay-Yong proposed translating *yi yu* as 'adding the areas formed by the yü', but she conceded that 'the word *fu* is prefixed to the names of the terms [···] in order to distinguish negative from positive coefficients' [Lam (1977), 265]. Lam Lay-Yong concluded that Yang Hui was familiar with negative terms [Lam (1977), 261]. This reading results in the aforementioned Problem 46 (Problem 9 in Guo Xihan)

---

[6]'四十八段田積內減三段不及步冪為實. 六之不及為從. 一虛隅,' or in modern transcription: $48A - 3a^2 + 6ax - x^2 = 0$.

[7]'和步冪內減十六之見積為實. 六之和步為從. 三步常法'.

[8]'益隅表示二次項系數為 -1' 246.

being rendered as $-x^2 + 60x = 864$.

In a description of two examples quoted by Yang Hui from the *Discussion on the Origin of Ancient Method*, however, Horiuchi [Horiuchi (2000), 246, 249] translated the term *yi*, which can also prefix yu, as 'added', which results in the 'added corner' or 'corner to be added', indicative of a positive quantity added to the dividend in the algorithm of resolution for the equation.[9] Consequently, the equality in the equation changes in the modern notation, $c + ax^2 = bx$. Horiuchi further noted that the equality suggested by the diagram is of the type $c = bx + ax^2$. For Problem 20 (Problem 62 in Lam Lay-Yong), Horiuchi read '$12sx = (4A - 12s^2) + x^2$', that is, an equation of the adjunct divisor ($12sx$) and the dividend ($4A - 12s^2$) to which a corner (x^2) is added. However, Horiuchi interpreted the accompanying diagram as $(4A - 12s^2) = 12sx - x^2$, but did not explain these two different expressions in detail. The discussion suggests that different readings of the equation are permitted and that the translation of *yi* as 'negative' is debatable. This reading calls into question the association of *yi* with *xu* Horiuchi did not explain the character *xu* for the simple reason that it never appeared in the extant part of the *Discussion on the Origin of Ancient Method*. Only 'augmented corner' *yiyu* 益隅, 'negative corner' *fuyu* 負隅, 'negative adjunct' *fu cong* 負從, and 'positive corner' *zhengyu* 正隅 appear in reference to positions on the tabular setting in *Yang Hui's Methods of Computation*. Apparently, only Li Ye uses the character *xu* in such situations.

The diagrams accompanying the descriptions of the Section of Pieces [of Areas] procedure are essential to understand the signs and equality. In order to consider negative terms, a negative geometrical area must be conceived and represented. With regard to the conceptualisation of a 'negative' piece of the field, a study of Problems 18 of the *Development of Pieces [of Areas]* and Problem 46 of *Yang Hui's Methods of Computation* offer illuminating examples. Problem 18 illustrates the use of *xu*, and Problem 46 clarifies the use of *yi*. On the basis of these examples, the character *xu* may be effectively translated as 'empty' and *yi* as 'augmented'.

---

[9][Horiuchi (2000), 249] : 'coin ajouté' or 'coin à ajouter'.

5.1   Xu, 'Empty' in the Section of Pieces of Areas

5.1.1   *Partial Translation*

Problem 18.

[18.1] Let us suppose that there is one piece of a circular field inside which there is a square pond. Outside the [area] occupied by water, one counts three hundred forty-seven *bu* of land. It is said only that the circumference of the outer circle and the perimeter of the inside square [added] together yields two hundred eight *bu*.
One asks how much are the outer circumference and inner perimeter each.
The answer: the circumference of the outer circle is one hundred eight *bu* and the perimeter of the inner square is one hundred *bu*.

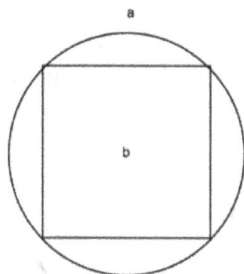

Fig. 5.1   a: 'circular field'. b: 'twenty-five *bu*'.

[18.10] One looks for this [i.e. the unknown] according to the Section of Pieces [of Areas] procedure. [From] the square of the *bu* of the sum, one subtracts twelve times the *bu* of the area to make the dividend. Eight times the *bu* of the mutual sum makes the empty adjunct (*xu cong* 虛從). Four is the constant divisor.

[18.11] The meaning is the following: Inside the twelve pieces of the circular field, there are twelve square ponds. Inside the square of the perimeter of the square, the twelve ponds are compensated once; outside, in what remains, four [ponds] are lacking. Therefore, with four, one makes the corner the divisor.

[18.12] The configuration (*shi* 式) originally empties the adjunct (*xu cong* 虛從). Now, on the contrary, the corner is emptied (*xu yu* 虛隅). That is why I recommend (*ming* 命) that four makes the empty constant divisor (*xu chang fa* 虛常法).

[18.13] The Old Procedure: self-multiply the *bu* of the mutual sum; place them in the top position. [Multiply] the *bu* of the area by

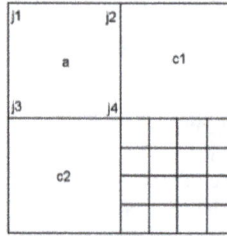

Fig. 5.2   a: 'This is the square of the circumference of the outer circle; it produces twelve areas of the circular field.' c1: 'Below are sixteen ponds; the side of the square makes four times the *bu* of the mutual sum. That is the adjunct.' c2: 'On the right are sixteen ponds; the side of the square makes four times the *bu* of the mutual sum. That is the adjunct.' j1–4: 'Subtract'.

twelve; subtract them from what is in the top position. Divide the remainder by eight to make the dividend. The *bu* of the mutual sum makes the adjunct. There is a divisor and an edge. The constant [divisor] is half a *bu*. Subtract the adjunct.

5.1.2   Description and Interpretation

[18.1] Let a be the sum of the circumference and the perimeter, 208 *bu* ('the *bu* of the sum' or 'the *bu* of the mutual sum'); let A be the area of the circular field (C) minus the area of the square pond (S), 347 *bu* ('the *bu* of the area'), and let x be the side of the pond.

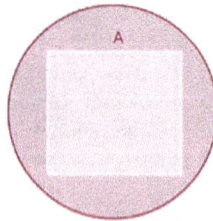

Fig. 5.3

This procedure requires the development of an explanation to understand the meanings of the paragraphs [18.10] to [18.12] above. The first sentence of the description of the Section of Pieces [of Areas] procedure lists the operations which generate the coefficients, as we have seen before. In this way [18.10], which reads '[From] the square of the *bu* of the sum, one subtracts twelve times the *bu* of the area to make the dividend (*shi* 實). Eight times the *bu* of the mutual sum makes the empty adjunct (*xu cong* 虛 從). Four is the constant divisor (*chang fa* 常法)' reported by the elements in [Table 5.1]

Table 5.1

$a^2 - 12A$	Dividend
$8a$	Empty adjunct
4	Constant divisor

Reading *xu cong* as 'negative adjunct' implies that the two other terms are positive. If the equality is made between a constant term and its indeterminate expression, this results in a transcription of the equation as $a^2 - 12A = -8ax + 4x^2$. This equation differs from the one presented in the Celestial Source procedure for the same problem [See Table A.1]. If the two procedures are considered equivalent, the transcription should read either $a^2 - 12A - 8ax + 4x^2 = 0$, like in the Celestial Source procedure, or $a^2 - 12A = 8ax - 4x^2$. Xu Yibao [Xu (1990), 66] transcribed the equation as $4x^2 - 8 \times 208x = 12 \times 347 - 208^2$ but did not compare this transcription to the one obtained by the Celestial Source procedure. The diagram and the 'meaning' [18.11] accompanying it suggest a better interpretation on the basis of the correlation of sentences with analysis of the diagram.

First, the figure is constructed using known data: a and A, the circumference of the field plus the perimeter of the pond and the area of the square field minus the area of the circular pond, respectively. With these two quantities, it is possible to represent a square of side a from which a square of $12A$ is removed [Fig. 5.4]. The result is the 'dividend' (*shi*) or the constant term of the equation. Next, the same area is expressed in terms of indeterminate quantities. The remaining dark area of [Fig. 5.4] is composed of two 'adjunct' rectangles of width $4x$ and length a. These two rectangles are superimposed on a square corresponding to $16x^2$. [Fig. 5.5] thus represents $8ax + 16x^2$. As in the previous problems, from 2 onwards, the superimposed area is displayed as a square external to the main figure [Fig. 5.6].[10] This extra square is removed. [Fig. 5.7] represents $8ax - 16x^2$. However, following this process, too much area is removed. Li

---

[10]See [Pollet (2012)].

Ye states that 'inside the twelve pieces of the circular field, there are twelve square ponds'. To produce the correct area, 12 squares of side x must be 'compensated', that is, added back to the original square and placed on the right side at the bottom. [Fig. 5.8] represents $8ax - 16x^2 + 12x^2$. Li Ye writes that 'inside the square of the perimeter of the square, once the twelve ponds were compensated, outside, in what remains, four [ponds] are lacking', namely, $8ax - 4x^2$. This area is equal to $a^2 - 12A$, as presented in [Fig. 5.4]. By following the procedure, the rectangles representing the 'adjunct' coefficient were literally 'emptied' in [Fig. 5.6] and [Fig. 5.7], while in the modern notation of the equation, $8ax$ remained positive.

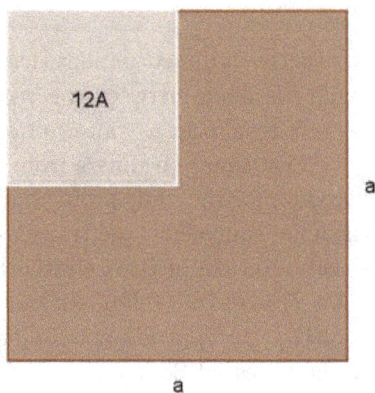

Fig. 5.4

The procedure above suggests the equation $a^2 - 12A = 8ax - 4x^2$. The 'dividend' ($a^2 - 12A$) is therefore positive, as is the 'adjunct' ($8ax$), but the 'corner' or the 'constant divisor' ($-4x^2$) is negative. The procedure in the diagram literally 'emptied the adjunct' rectangles; that is, it removed the extra area made by the superimposition of adjunct rectangles. This representation is why xu is translated as 'empty' or 'to empty'. Thus, the modern concept of 'negative' is distinct from the word 'empty.' The word xu denotes the removal of an area, as opposed to other areas, which are 'full'. On the other hand, the 'corner' was filled. These details demonstrate the caution needed in transcribing equations into modern symbolic language: here, xu does not name a sign but a relation between signs.

More philosophically, the practitioner seemingly operates according to

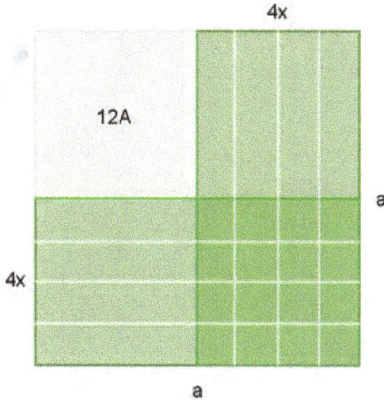

Fig. 5.5

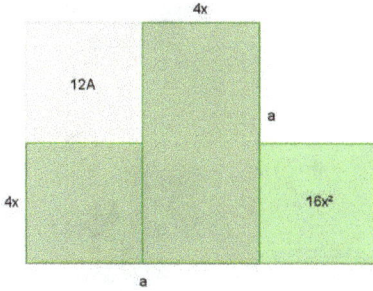

Fig. 5.6

a binary system. This system is also a source of movement, where void has a positive and constitutive value. An area, but more generally any object, is identified by its contrary, not by the identity to itself. There is a change of algebraic sign where values are inverted, where contraries define objects. Movement is not only a matter of transformation of shape, but also of areas.

In [18.12] Li Ye adds a recommendation that may be understood in this context. 'The configuration originally empties the adjunct. Now, on the contrary, the corner is emptied. That is why I recommend that four makes

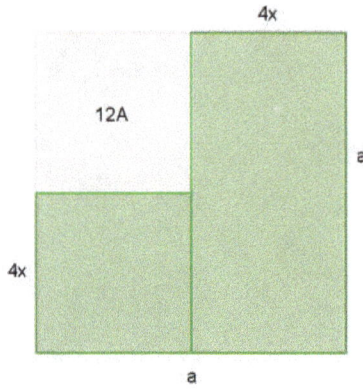

Fig. 5.7

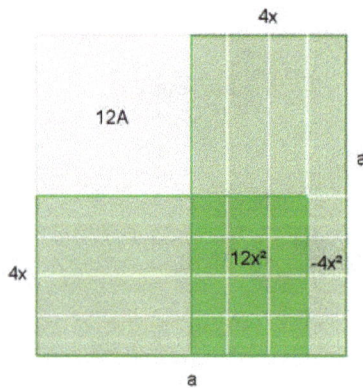

Fig. 5.8

the empty constant divisor'.[11] In the procedure above, the adjunct was 'emptied'. After that, the lost area was restored by adding twelve squares of side x. In a short sentence, Li Ye recommended emptying the corner instead of the adjunct rectangle. Instead of displaying the extra square area in [Fig. 5.6], removing it and adding 12 squares in compensation, Li Ye recommends directly performing the transformations on the 16 squares

---

[11]The same type of recommendation happens in Problem14.

of side x [bottom right on Fig. 5.5]. From these 16 squares, 4 squares are removed. Instead of $8ax + 16x^2 - 12x^2$, the recommendation of Li Ye directly generates $8ax - 4x^2$. In other words, the steps in [Fig. 5.6] and [Fig. 5.7] are skipped. Thus, the extra squares of the adjunct need not be removed and the lost area need not be compensated, because the 'corner' (*yu* 隅) was 'emptied' (*xu* 虛). In this way, Li Ye improves the computational economy of the procedure.

Concerning this example, there are two conclusions. First, the term *xu* refers to the transformation of areas into diagrams and is extended to include the entries on the counting surface corresponding to these areas. This term only applies to coefficients of indeterminate terms and not to the dividend, which is always positive. An area can be *xu* while its coefficient remains positive. Eventually, *xu* was generalised to name an expression of the area in terms of the indeterminate in the Celestial Source procedure, as opposed to its expression in terms of the constant. This opposition may explain the use of *xu* in the procedure of the Celestial Source in Problems 1 and 2 mentioned in the introduction of this chapter [See footnote 3]. Second, in addition to the two geometrical procedures (the Section of Pieces [of Areas] procedure and the Old Procedure), a third type of procedure survives in a commentary on the Section of Pieces [of Areas] procedure: the 'original' (*yuan* 元) that is to be distinguished from the 'recommended' (*ming* 命) by Li Ye. Therefore, there is more than one way to read the procedure in the *Development of Pieces [of Areas]*. Before this point can be addressed, though, the Old Procedure must be analysed, and this procedure depends on an understanding of the *yi yu*, in *Yang Hui's Methods of Computation*. A translation and mathematical description of Problem 46 of *Yang Hui's Methods of Computation* illustrates the meanings of these terms.

5.2 Yi, 'Augmented', in Problem 46 of *Yang Hui's Methods of Computation*

The translation and transcription follow the format of Problem 18 of the *Development of Pieces [of Areas]*.

5.2.1 *Translation*

Problem 46.

[46.1] The area of a rectangular field is eight hundred sixty-four *bu*. It is said only that the length and the breadth together are sixty *bu*.

Look first for the *bu* of the breadth.
The answer: twenty-four *bu*.

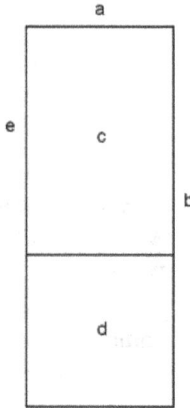

Fig. 5.9 a: 闊二十四, 'Breadth: Twenty-Four'; b: 一長一闊共六十步為從方, 'Sum of One Length and One Breadth: Sixty *bu* as Adjunct Square'; c: 本積八百六十四步, 'Original Area: Eight-Hundred Sixty-Four *bu*'; d: 益闊方積五百七十六, 'Augmented Area of the Square of the Breadth: Five-Hundred Seventy-Six'; e: 長三十六, 'Length: Thirty-Six'.

[46.2] The method of the augmented corner: the area is placed to make the dividend. The *bu* of the sum makes the adjunct square, one makes the augmented corner. Open the square by dividing.

[46.3] The *Development of Pieces [of Areas]*: one area has only one length. If the *bu* of the sum of the length and breadth makes the adjunct square, [the area] is short of one breadth. Therefore, one [square] is used to make the augmented corner, adding one piece of the square of the breadth. [...] [12]

5.2.2 *Description and Interpretation*

[46.1] Let A be the area of the rectangle, L be the length, a be the sum of the length and breadth and x be the breadth.

As in the *Development of Pieces [of Areas]*, the description in [46.2] can be summarised in [Table 5.2].

---

[12] [46.1] 真田積八百六十四步, 只云長闊共六十步. 卻先求闊步得几何. 答曰: 二十四步. [46.2] 益隅術曰: 置積為實, 共步為從方, 以一為益隅, 開平方除之. [46.3] 演段曰: 一積止有一長. 若以長闊共步為從方, 正少一闊, 所以用一位益隅, 入一段闊方, 以應從方除數. [...]

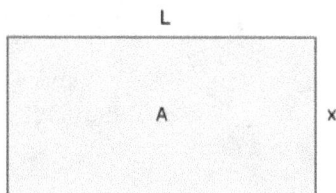

Fig. 5.10

Table 5.2

A	Dividend (*shi* 實)
a	Adjunct Square (*cong fang* 從方)
1	Augmented Corner (*yi yu* 益隅)

In [46.3], A is 'the area placed to make the dividend'. 'The *bu* of the sum', a, 'makes the adjunct square' and 'the augmented corner' is one.

Next, a diagram is constructed with the known data. Because A is known, a rectangle with area equal to A is drawn. Now, a superimposed rectangle of length a, the sum of the breadth and width, must be drawn and the two rectangles must fit together. However, the segment a is too long to fit with the length of the rectangle of area A. Thus, the area A lacks some quantity, which Yang Hui describes as 'one area has only one length. If the *bu* of the sum of the length and breadth makes the adjunct square, [the area] is short of one breadth'. Now, if a rectangle of length a and breadth x is compared to A, there is an extra square of side x inside the adjunct rectangle, ax. Therefore, in [Fig. 5.11], 'one [square] is used to make the augmented corner, adding one piece of the square of the breadth'. In this way, a corner (*yu* 隅) augments (*yi* 益) the area of A. The diagram is thus read as $A + x^2 = ax$.

Yi and *xu* can both be explained geometrically, but they do not have the same origin. *Yi* is applied to a supplement of the surface, while *xu* corresponds to an area which is taken out. Therefore, *yi* and *xu* are translated differently as 'augmented' and 'emptied'. The former is an added figure and the latter is an emptied figure. An analysis of the texts shows that a negative coefficient cannot be directly expressed. The transformations implied by the text, however, read as if the author may have considered negative coefficients. Thus the conceptualisation found in the *Collection Augmenting the Ancient [knowledge]* and the *Discussion on the Origin of Ancient*

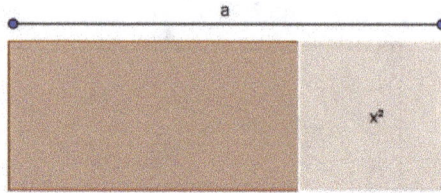

Fig. 5.11

Method must be clearly distinguished from the mathematical concepts of Li Ye and Yang Hui which appear in the next chapter. The concept of negative coefficients was apparently not clear and the expression of equality is ambiguous. There are different ways to read the two layers of diagrams. In the *Development of Pieces [of Areas]*, the diagram is always read as a known area, re-expressed in terms of unknowns and is transcribed as $ax + bx^2 = c$. In *Yang Hui's Methods of Computation*, a diagram can be read in different ways and transcribed as either $ax + bx^2 = c$ or $c + ax^2 = bx$, depending on the problem. Yang Hui does not use the diagrams as systematically as Li Ye.

These two examples illustrate another important difference between the two treatises. In the *Development of Pieces [of Areas]*, the number of adjunct rectangles is always even but no more than one adjunct rectangle appears in *Yang Hui's Methods of Computation*. A consideration of the Old Procedure in the *Development of Pieces [of Areas]* explains this phenomenon.

Chapter 6

The Old Procedure: The Example of Problem 5

6.1 Diagram of the Old Procedure

Twenty-three of the problems of the *Development of Pieces [of Areas]* refer to the Old Procedure.[1] This procedure is described in one sentence, similar to the sentence opening the procedure of the Section of Pieces [of Areas] which also describes the computation of the coefficients of the equation. As was the case with the Sections of Pieces [of Areas] procedure, this passage describes a manipulation performed on the counting surface. The top position, where the 'dividend' is constructed, is mentioned in many of the problems,[2] and some coefficients are moved 'backwards' (Problems 9, 10), 'to the left' (Problem 44), or shifted to the lower rank (Problem 45).

This procedure is also geometrical and involves drawing a diagram, like in the Section of Pieces [of Areas] procedure. Nonetheless, Problem 22 is the only case with a surviving diagram.[3] A few comments by Li Ye help to clarify some of the terms in the equations by means of geometry and sometimes clearly indicate that drawing the diagram constitutes part of the solution:

'The Old Procedure and the new one are different. The Old Procedure is always simple. This mathematical procedure favours only what is easy and simple. It has been replaced by a new procedure because the Section of Pieces [of Areas] is difficult to draw with the Old Procedure. [This new procedure] is like [the old one] and complements it.'[4] In contrast to the

---

[1]Problems 5, 6a, 6c, 8, 9, 10, 13, 14, 15, 16, 18, 19, 21, 22, 33, 44, 45, 46, 51, 56, 57, 59, 60 and 62.

[2]Problems 5, 8, 9, 10, 13, 14, 15, 18, 19, 22, 33, 45, 46, 57 and 62.

[3]Problem 56 also contains two diagrams for the Old Procedure, but those diagrams were added by the editor of the Qing Dynasty imperial encyclopaedia, the *Complete Library*.

[4]Problem 15: '新舊二術不同者. 舊術從簡耳. 算術本貴簡易而猶立新術者, 緣舊術難畫條段也. 餘倣此.' This commentary will make sense later.

Section of Pieces [of Areas] procedure described in examples before, there
are no characters *xu, yi, fu* or *zheng* for negative quantities. Nowhere in the
discourse of the Old Procedure are there any expressions for indeterminate
terms, equality or signs.

The Old Procedure is usually stated in a single paragraph. For example,
the Old Procedure for Problem 18 is stated in [18.13]: '*The Old Procedure:
self-multiply the bu of the mutual sum; place them in the top position.
[Multiply] the bu of the area by twelve; subtract them from what is in the
top position. Divide the remainder by eight to make the dividend. The bu
of the mutual sum makes the adjunct. There is a divisor and an edge. The
constant [divisor] is half a bu. Subtract the adjunct.*'

The example of Problem 18 discussed above presents a case of the division of coefficients by a common factor, in which the coefficients of the
equation stated in the procedure of the Section of Pieces [of Areas] have
been divided by 8. The given values may be tabulated as [Table 6.1]

Table 6.1

$\frac{a^2-12A}{8}$	Dividend
a	Adjunct
0.5	Constant Divisor

From [18.13] and what is known about the Section of Pieces [of Areas]
procedure, a tentative reconstruction of the Old Procedure may be proposed. First, a figure is constructed with the known values for a and A,
as in previous cases in [Fig. 5.4]. Second, the same area is expressed in
terms of the indeterminate [Fig. 5.5]. Each of the two rectangles of [Fig.
5.5] are composed of four bands of width x and length a. In fact, the total
of eight bands can be assembled like a long ribbon of length $8a$ [Fig. 6.1].
Following the information given for the dividend above, this ribbon should
be divided by 8. To reword the figure in terms of an unknown variable,
[Fig. 6.2] represents the result (ax) of the division of the adjunct rectangle.
[Fig. 6.2] includes four squares of side x. In the previous procedure, the
term $4x^2$ had to be removed. For the same reason, $\frac{4x^2}{8}$ must be removed
here. [Fig. 6.3] represents $ax - 0.5x^2$. This rectangle is equal to $a^2 - 12A$
divided by eight. Therefore, the equation becomes $\frac{a^2-12A}{8} = ax - 0.5x^2$.

Here, the multiple adjunct rectangles of [Fig. 5.5] are reduced to a
single rectangle. Indeed, in twelve of the cases of the Old Procedure, there
is no more than a single 'adjunct', that is, either $x = 0$ or the 'adjunct' is

Fig. 6.1

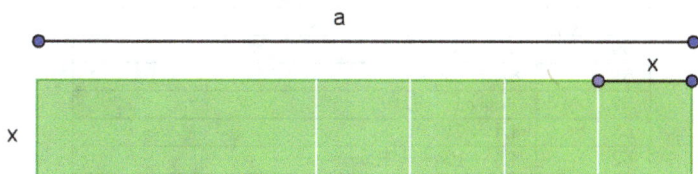

Fig. 6.2

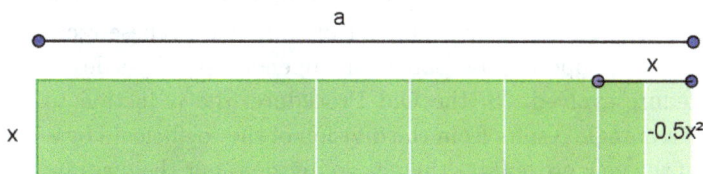

Fig. 6.3

transcribed as ax in [Table A.1].[5] The same pattern is observed in Problem 46 of *Yang Hui's Methods of Computation* translated above, and other problems in this work share the same pattern.[6]

The patterns of the Old Procedure of the *Development of Pieces [of Areas]* and *Yang Hui's Methods of Computation* resemble each other. *Yang Hui's Methods of Computation*, like the Old Procedure of Problem 18 of the *Development of Pieces [of Areas]*, reduces the 'adjunct' rectangles to a single rectangle. [Table 6.2] collects the equations of *Yang Hui's Methods of Computation* for which there is either no adjunct $(x = 0)$ or a single

---

[5]Problems 6a, 6b, 8, 15, 16, 18, 21, 33, 45, 51, 59 and 60.
[6]Problems 49, 50, 51, 52 and 53.

Table 6.2   Equations of *Yang Hui's Methods of Computation*

Problem	Equation
43	$A = ax - x^2$
44	$A = x^2 - ax$
45	$4A + a^2 = x$
46	$A = ax - x^2$
47	$A = ax - x^2$
48	$a^2 - 4A = x^2$
49	$3A = ax - 5x^2$
50	$5A = ax - 3x^2$
51	$8A = ax - x^2$
52	$A = ax - 8x^2$
53	$4A = 7x^2$
62	$4A - 12a^2 = 12ax - x^2$
63	$12(a^2 + A) = x^2$
64	$4A - 3a^2 = 4x^2$ corrected to $3a^2 - 4A = 4x^2$

adjunct (ax).[7] In the cases presented in the descriptions of the Section of Pieces [of Areas] procedure of the *Development of Pieces [of Areas]* (i.e. the 'original' or 'recommended' procedure of Problem 18), the adjunct is always represented by an even number of rectangles. These rectangles are either multiple, superimposed areas or appended side-by-side. Therefore, the procedure evolved. In the Old Procedure, the reduction to a single 'adjunct' rectangle results from the division of the coefficients by a common factor. In the new procedures there is no division, but the diagram presents multiple adjunct rectangles.

The change of procedure affects the drawing of the diagram. In the Old Procedure of Problem 18, several diagrams of different shapes are necessary, whereas the Section of Pieces [of Areas] procedure uses only one diagram for the same problem. This detail might explain why Li Ye differentiates between the two procedures in Problem 15 quoted in the introduction of this chapter. Indeed, in Problem 18 and other similar cases, the Old Procedure is easier because the adjunct coefficient is reduced to a single piece in the diagram and to smaller quantities on the counting surface. Conversely, the diagrams are more complex because they require several drawings. The new procedure of the Section of Pieces [of Areas] employs more complex and larger quantities for 'adjunct' rectangles on the counting support but the drawing remains simple. A single diagram is sufficient to illustrate the whole procedure.

---

[7] There is an exception in *Yang Hui's Methods of Computation* in Problem 62, which is described in Xu Yibao (1990).

6.2   Computing Coefficients with the Old Procedure

Further analysis of the Old Procedure yields more new information on the evolution of the procedure and its writing. This evolution also occurs through the medium of manipulations on the counting surface and the ways of expressing these manipulations textually. That is, the algorithm and its narration changed from one procedure to the other. These changes are reflected in [Table A.1], which reports the following:

(1) In eighteen cases, the Old Procedure presents the same equation as the Section of Pieces [of Areas] procedure with the difference that all the coefficients are divided by common factors.
(2) In six cases (Problems 5, 10, 13, 14, 46 and 57), the Old Procedure presents an equation in which the coefficients are factored by three.
(3) In one case, the Old Procedure generates an equation with a negative dividend (Problem 46).

The cases listed in (1) correspond to the evolution of the diagram illustrated by Problem 18 previously. There are other differences that suggest other paths of evolution for the same procedure. The other cases that do not involve division (Problems 5, 10, 13, 14, 46 and 57) reveal other fine points for each type of procedure. Problem 5 of the *Development of Pieces [of Areas]* constitutes a representative case study of this phenomenon. Problem 5 exhibits differences not only between the diagrams of the Old Procedure and the Section by Pieces [of Areas] procedure but also between the algorithms and manipulations on the counting surface. Problem 5 states:

Problem 5.

[5.1] Let us suppose that there is one square piece of field, inside which there is a circular pond. Outside the [area] occupied by water, one counts thirteen mu two fen of land. It is said only that the circumference of the inner circle does not attain the perimeter of the outer square by one hundred sixty-eight *bu*. One asks how long the circumference [⋯]
Let a be the difference between the perimeter (p) of the square and the circumference (c) of the pond, 168 *bu* (the *bu* that does not attain). Let A be the area of the square field (S) minus the area of the circular pond (C), 13 *mu* 2 *fen* or 3168 *bu* (1 *mu* = 240 *bu*). Let x be the circumference.

Concerning the procedure of the Section of Pieces [of Areas], paragraph [5.10] states: '*from forty-eight pieces of area of the field, subtract three pieces*

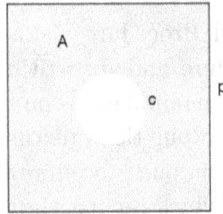

Fig. 6.4

of the square of the *bu*, which does not attain to make the dividend. *Six times the [bu] which does not attain makes the adjunct. The empty corner is one*', which can be tabulated as [Table 6.3].

Table 6.3

$48A - 3a^2$	Dividend
$6a$	Adjunct
1	Empty corner

The Old Procedure reads as follows [5.12]: '*the area of the field multiplied by sixteen makes what is in the top position. Self-multiply the bu which does not attain, subtract them from what is on the top position and triple what remains to make the dividend. Six times the bu which does not attain makes the adjunct. There is a divisor and an edge. One bu is taken to make the constant [divisor]*', which can be set in [Table 6.4].

Table 6.4

$3(16A - a^2)$	Dividend
	Divisor-edge (*Fa-Lian* 法廉)
$6a$	Adjunct
1	Constant [divisor]

From these values, the Section of Pieces [of Areas] generates the equation $48A - 3a^2 = 6ax - x^2$ and the Old Procedure produces $3(16A - a^2) = 6ax - x^2$.

According to these elements, there are two differences between the procedures. The first difference appears in the construction of the dividend (treated in 1) and the second difference occurs in the names of the rows on the counting surface (treated in 2). The operations for computing the dividend are stated differently in the two procedures. Both procedures employ multiplication by three, but three appears as a factor in only the Old Procedure. Apart from this difference, the Old Procedure and the Section of Pieces [of Areas] procedure present the same equation. Why record two procedures if they are almost the same? The second difference is the specific vocabulary of the Old Procedure, which differs from the Section of Pieces [of Areas] procedure. The 'dividend' is still named *shi*; the 'adjunct' is still called *cong*. 'Constant', *chang*, still refers to the last row in which the coefficient of the second-order indeterminate term is constructed. However, two new characters appear: divisor (*fa* 法) and edge (*lian* 廉).[8] These two characters never appear together in the Section of Pieces [of Areas], which calls the adjunct '*cong*'. *Cong fa* and *lian fa* are in fact appellations found in older texts.

6.2.1   *Multiplication by Three.*

The different ways of stating the coefficient imply different computations. For instance in Problem 5, a multiplication by 48 is required. Yet, in each of the procedures this operation is expressed differently implying different executions. In the Section of Pieces [of Areas] procedure of Problem 5, as before, A is the 'area of the field' (that is, the square area minus a circular area) and a is the '*bu* that does not attain' (the difference between the circumference and the perimeter). The dividend can be written as $48A - 3a^2$. In the Old Procedure, it becomes $(16A-a^2)\times 3$. In the Celestial Source procedure, 48 is directly involved in the computation at the beginning of the problem. Li Ye states the 48 is 'a denominator' which directly multiplies every quantity. In the different procedures, the multiplication by three is not treated in the same way. How can various roles of this multiplication by three be explained?

Usually when computing the area of a circle in the *Development of Pieces [of Areas]*, the diameter is multiplied by three and then divided by four. The *Development of Pieces [of Areas]* thus approximates the value of π as three. Arguably, this value does not mean that the author was not aware of the fact that π is not equal to three. Rather, the value is the result of a process which calculates the areas of circular figures from corresponding square figures. Thus, the four circular areas are equivalent to three squares

---

[8]Problems 5, 8, 9, 10, 13, 14, 15, 18, 19, 22, 46, 51 and 57.

with sides equal to their diameter. This arithmetical transformation is convenient for the geometrical transformation of areas. To represent four circles, three squares may be drawn instead, as seen in Transformation C.

In general, in the Celestial Source procedure, this process appears as prescribed multiplications and divisions. The algorithms of multiplication and division reflect the fact that the operations are inverses.[9] To perform a multiplication on a counting surface, the multiplier is placed in the upper position and the multiplicand fills the lower position. The multiplicand moves to the left according to the number of digits. Similar to multiplication, wherein operations relate to the position of the multiplier relative to the multiplicand, the operations are based on placing the divisor relative to the dividend in division. The quotient (*shang* 商) is placed on top. The result is represented by a mixed fraction $A\frac{b}{c}$, where A is in the 'quotient' row, b is called the 'numerator' (*zi* 子) and c is called the 'denominator' (*mu* 母). The denominator shares a row with the divisor [see Appendix A.2].

In the Celestial Source procedure, it is sometimes more convenient to transform circles into squares by multiplying by 3 without dividing by 4, because 4 circular areas are $\frac{4 \times d^2}{4} = 3d^2$, or 3 squares of side d. The number 4 is supposed to be a divisor and it probably remains on the counting surface at the position of denominator. The number 4 is a kind of 'remainder'.[10] Although intended as a divisor, it is placed in the denominator position. The position as the multiplier in multiplication is the same as the divisor in division. Later, in the computation of the other area, this denominator is used as a multiplier, by which the area of the square is multiplied. Thus, one finally obtains the areas of 3 squares and 4 circles, which can be subtracted from one another in the Celestial Source procedure.

Problem 5 presents a slightly different situation. Here, the denominator is 48 and not 4. 48 is a multiple of both 3 and 4, and the square considered in the first part of the procedure of the Celestial Source [5.2] and [5.3] has four sides equal to the circumference. The circumference (c), the diameter (d) and the area of the circle (C) are related as $c^2 = 9(d^2) = 12C$. Yet, at the beginning of the procedure, only the square of the perimeter is considered as $p^2 = 16S$. $12C$ and $16S$ cannot be cancelled. The operation that is usually

---

[9][Li and Du (1987), 15–19]; [Chemla and Guo (2004), 15–10]; [Lam and Ang (2004), ch. 3–4]; [Martzloff (1987), 229–249)], among others.

[10]This process recurs in several problems where denominators are used as multiplicands. For example, in Problem 11a, Li Ye wrote, 'with the denominator, quadrupling this [⋯]', '就分母四之 [⋯]'

applied to the side and the diameter is transferred to the perimeter and the circumference. The multiplication of the square of the perimeter by three received the comment that 'the reason for the multiplication by three and for making of forty-eight is that it makes forty-eight for denominator' in the Celestial Source procedure. The square of circumference is multiplied by four, and 48 is set in the position of the 'denominator' and used to multiply every area. This set of operations appears differently in the Section of Pieces [of Areas] and in the Old Procedure. This disparity reveals different mathematical executions.

In the Section of Pieces [of Areas] of [5.10], 48 is directly reported as the multiplicand of the area with no reference to any fraction. Yet, in the 'meaning' of the procedure [5.11], the coefficient 48 is constructed. The first square of side d is considered and subsequently, its area is tripled. The multiplication by four is never stated, although it is expressed by the removal of c^2 three times from the rectangle of the adjunct and one time more from the outside square, which constitutes four removals. These multiplications by three and four are the sole elements of the procedure preserved in the discourse [5.11]. The remaining procedure, by which the coefficient is constructed and the multiplications are justified, appears in the diagram. The final result of the computation with rods and drawing diagrams is the conclusion to [5.10], which states the coefficients, placed clearly on the surface, are ready for the extraction of the root. That is, only when the manipulation of rods and the drawing have already been performed, the equation is finally explicitly written.

In the Old Procedure, instead of directly reporting $48A$, $16A$ is set out first. However, the equation is not yet ready for the extraction of the root. The computation must be adjusted to find the correct dividend. This is done for one square of side p, where $p^2 = 16S$, 'in order to fit with the area of [the square] of the perimeter of the square' as Li Ye states in his commentary. $16A$ is mentioned in [5.12]. $16S$ can be drawn, $16A$ can be computed, but $16C$ is not given. So the equivalence $16A = 16S - 16C$ is not possible because $16C$ is not known. However, the fact that $c^2 = 12C$ is known, and it is possible to derive $16A$ from $12C$ by multiplying them by three and then by four, but the multiplication by three appears later in the algorithm than in the Section of Pieces [of Areas] procedure. It appears as if a correction was made after the coefficients were found.

The multiplication by three, π, occurs at different steps in each of the three procedures. In the Old Procedure, the coefficients are adjusted after the multiplication, requiring a few more steps before the extraction of the root. In the Section of Pieces [of Areas] procedure, the multiplication is immediately stated in the discourse and the diagram and is integrated into the construction of the coefficients. At the end of the procedure, the lay-out is ready for the extraction of the root. In the Celestial Source procedure, 48 is immediately multiplied by each expression of the areas at the very beginning. The difference among the statements of the procedures reflects the differences in the execution of the algorithms. The evolution of the relation between the algorithm and its discourse indicates the evolution of the status of the operation. First used as a correction, the operation became a presupposition later.

This multiplication by three, π, is indeed well known in *Yang Hui's Methods of Computation.* Problem 53 invokes the claim that 'the area of four circular fields equals that of three square fields'.[11] The sequence of operations is similar to that of the Section of Pieces [of Areas] procedure. The multiplication by three in *Yang Hui's Methods of Computation* resembles the Section of Pieces [of Areas] in the *Development of Pieces [of Areas]*, while the use of the adjunct corresponds to that of the Old Procedure.

For a better understanding of the various procedures, a study of the differences with Problem 5 of *Development of Pieces [of Areas]* is a useful object lesson. As noted above, the Old Procedure and the Section of Pieces [of Areas] procedure also differ in the names of their rows. The Old Procedure of the *Development of Pieces [of Areas]* contains a 'divisor' and an 'edge'.

6.2.2 *Counting Support Row and Signs of Coefficients*

The rows called *lian* and *fa* also appear in *Yang Hui's Methods of Computation* with a clear description of their roles. These two names refer to the same row at different moments in the algorithm of root extraction. These rows appear in the section titled 'working', not in the one titled '*yan duan*'. This row, which already appeared in the algorithm of division, holds the intermediate quantities required for computation. According to the algorithm reconstituted by Lam Lay-Yong [Lam (1977)] and Guo Shuchun [Guo (1991)], the extraction of the root follows several steps. First, the hundreds of the root are computed on the *fang fa* row, and second, the tens are computed on the *lian* row.

In this procedure of root extraction, *fang fa* and *lian* are names of a row inspired by a geometrical representation of the root extraction, as has been

---

[11]'而四圓田積及三個方田'.

made clear by historians.[12] No diagram presenting the root extraction is given by Yang Hui after this procedure. Yet, in a similar problem, a second procedure, called 'subtract the adjunct' (*jian cong* 減從) is provided with a diagram. This procedure differs slightly from the one given above. The same operations are involved, but their order is different. The important difference between the two procedures concerns the geometrical approach: the two diagrams are transformed in completely different ways [Lam (1977), 260–262]. There is also another difference: the second procedure has no *fang fa* or *lian* rows. In this case, the operations are done on the row of the adjunct and the quantity of the adjunct is diminished step-by-step.[13] The specific procedure is chosen according to the signs of the coefficients. If the corner or the adjunct is negative, the procedures called 'augmented corner' (*yi yu* 益隅) or 'subtract the adjunct' (*jian cong* 減從) are prescribed. If the terms are positive, a different procedure named 'to extract the root of

---

[12][Lam (1977), 248]: '[It] is more probable that the methods were first conceived from the diagrams and the steps of the working were deduced, rather than vice versa'.

[13]In *Yang Hui's Methods of Computation*, the *fang fa* and *lian* rows appear in only Problems 43, 44, 46 and 53 for this type of procedure. They never appear in the procedure of *jian cong*.

Problem 46 presents an equation of the type $A = ax - x^2$, and two methods to extract its root. The procedures are given without illustration but with references to the counting surface. The first procedure reads:

'The working: Place the area eight-hundred-sixty-four *bu* as the dividend (*shi* 實). [Step 1] Place one counting rod separately as the augmented corner (*yi yu* 益隅). [Step 2] From the position of last digit of the dividend, [move this] to reach the digit of the hundredth [on the row] below to determine the tenth. Put twenty as the breadth in the row of the above quotient (*shang shang* 上商). [Step 3] Below the area [i.e. the dividend], place the square divisor (*fang fa* 方法), twenty. [Step 4] The above quotient calls (*ming* 命) the square divisor and multiplies it. It is four hundred, the augmented area. [Step 5] Next, [multiply the square divisor by] the adjunct divisor, sixty *bu*. The area is one thousand two hundred *bu*. [There] remain sixty-four *bu*. [Step 6] Multiply the square divisor by two, move it back from one place to make the edge (*lian* 廉). The adjunct divisor is also moved back by one place. The augmented corner is moved back by two places. [Step 7] [Place a] further four *bu* in the above quotient. Next to the edge, place the corner, four. Multiply the edge corner by the above quotient. [Step 8] Added to the augmented area in the dividend is two hundred and forty. [Step 9] The above quotient calls the adjunct divisor. What remains at the dividend is exhausted. It yields a breadth of twenty-four *bu*. Hence, [this] is the answer.'

'草曰: 置積八百六四步為實. 別置一算為益隅. 從尾末位約實至百下定十. 上商闊二十. 積下置方法二十, 以上商命方法, 乘四百益積. 卻, 以從方六十, 除積一千二百, 餘六十四. 二因方法, 一退為廉. 從方亦一退. 益隅再二退. 又上商闊四步. 廉次之下亦置隅四. 以上商乘廉, 隅, 益積實共二百四十. 上商命從法, 除實盡. 得闊二十四步. 何問.'

The captions [Step 1] to [Step 9] were added to coordinate with the table following the translation. The steps may be pictured in reference to the counting support. Lam Lay-Yong [Lam (1977)] and Guo Xihan [Guo (1996)] already provide details concerning the procedures.

the square plus the adjunct' (*dai cong kai fang* 帶從開方) is prescribed.

According to Li Ye, the *jian cong* procedure is supposed to be used for Problem 5. Li Ye indicates this procedure for cases wherein the constant divisor is negative; otherwise no specific instructions are given.[14] In [Table A.1], among the equations of the Old Procedure that have been transcribed into modern notation, the adjunct are always positive. Therefore, in the Old Procedure, only the constant divisor can be negative.

In contrast to Yang Hui's observation on the Old Procedure of the *Development of Pieces [of Areas]*, the *fang fa* and *lian* rows appear regardless of the procedure chosen for the root extraction. Li Ye states in Problem 14 that these rows constitute the main difference between the Old Procedure and the Section of Pieces [of Areas] procedure: 'In the new and the old [procedures], the edge and the adjunct are not the same thing. Yet, when opening [the square], [the procedure] is the same. Therefore, these two [procedures] are to be preserved'. [15] However, Li Ye never details the procedure of extracting roots in the *Development of Pieces [of Areas]*. At the time of Li Ye, the resemblance between the procedures 'augmented corner' and 'subtract the adjunct' was perfectly clear and diagrams were not even required. That is why he states here that '[the procedure] is the same'. However, a *fa-lian* row is preserved here either as a vestige of more ancient practices or as a reconstitution by Li Ye on the basis of what he supposed to be ancient. In the Old Procedure, operations are done on this row; in the new one, they are done on the adjunct row directly, regardless of the signs of the coefficients. Indeed, between the moment when the signs were

	Step 1	Step 2	Step 3	Step 4	Step 5
Shang Shang			2	2	2
Shi	8 6 4	8 6 4	8 6 4	1 2 6 4	6 4
Fang Fa			2 0	2 0	2 0
Cong	6 0	6 0	6 0	6 0	6 0
Yi Yu	1	1	1	1	1
Explanation			$20 \times 20 = 400$ $864 + 400 = 1264$	$20 \times 60 = 1200$ $1264 - 1200 = 64$	

	Step 6	Step 7	Step 8	Step 9
Shang Shang	2	2 4	2 4	2 4
Shi	6 4	6 4	2 4 0	2 4
Lian	4 0	4 4	4 4	4 4
Cong	6 0	6 0	6 0	6 0
Yi Yu	1	1	1	1
Explanation	$20 \times 2 = 40$	$40 + 4 = 44$ $44 \times 4 = 176$	$176 + 64 = 240$	$60 \times 4 = 240$ $240 - 240 = 0$

[14]'subtract the adjunct' is in the Old Procedure of problem 5, 14, 18, 46, 51 and 57.
[15]'新舊廉從不同. 開時, 則同. 故兩存之.'

geometrically considered in the Section of Pieces [of Areas] procedure or in the Old Procedure, and the time when Li Ye recorded these two procedures, things changed. The study of the sign of the dividend is interesting on this point.

Negative constant terms seem to be impossible in geometrical procedures. Lam Lay-Yong [Lam (1984), 260] noticed that the constant term (or dividend) in the Section of Pieces [of Areas] procedure is always positive. This term results from a subtraction of the area given in the statement from one or several squares of the segment given in the statement. The smaller number is always subtracted from the larger. However, for example, in the Celestial Source procedure for Problems 1 and 2, the dividend is transcribed as $A - 4a^2$, where A is the area and a is a segment given in the statement of the problem. In Problem 1, this term is positive, while in Problem 2, the term is negative. In the Section of Pieces [of Areas] procedure for Problem 1, the dividend is expressed as $A - 4a^2$ and this term is still positive, but in Problem 2, it is expressed as $4a^2 - A$. This change renders the dividend positive and repeatedly recurs in the *Development of Pieces [of Areas]*.

Surprisingly, the equation for the Old Procedure of Problem 46 in the *Development of Pieces [of Areas]* contains a negative dividend. However, in the Section of Pieces [of Areas] procedure, it is positive. The Celestial Source procedure presents a negative dividend as well [Table A.1 in Appendix A.1]. Unfortunately, Li Ye does not comment upon this problem. Concerning the signs of the coefficients, he comments on problem 14 that '[whenever there is] a negative adjunct with a positive corner, or a positive adjunct with a negative corner, the dividend always amounts to the same. Therefore, that is the reason why the edge and the adjunct are used to differentiate.' Li Ye seemingly considers that the dividend has no sign, or that its sign is not to be taken into account, unlike the signs of the other coefficients. The dividend is always constructed through a subtraction. Indeed, all the subtractions performed to construct the dividend in *Yang Hui's Methods of Computation* or the *Development of Pieces [of Areas]* are done with the goal of obtaining a positive result because of geometrical constraints.[16] In the Celestial Source procedure, which does not require geometry, the subtraction is done in a random 'direction', and the dividend can either be negative or positive. Li Ye does not consider the

---

[16] *Yang Hui's Methods of Computation* contains one problem with a case of a negative dividend. Yang Hui comments that 'the ancients said that this method is awkward', '古人謂法拙.' He corrected 'The smaller is subtracted from the bigger', '以少減多' to obtain a positive quantity.

sign of the constant coefficient important; only its 'absolute value' appears in the root extraction, because it is sufficient for any procedure for root extraction. Yet, the signs of the other terms (i.e. x and x^2) determine the particular procedure for the root extraction. The procedure varies according to whether these coefficients are positive or negative and whether the old or new procedures are used, as we saw previously. The variability of the procedure may justify the absence of any commentary to Problem 46 of the *Development of Pieces [of Areas]*. For ancient mathematicians, signs were an obligation imposed by geometry; this ceased to be a necessity during the time of Li Ye.

This remark, as well as the comment to Problem 14 by Li Ye, reveals one last observation. The vocabulary used for negative coefficients and reference to diagrams in the commentary differs from that of the discourse of the procedure. Li Ye wrote about *fu* and *zheng* coefficients, whereas the Section of Pieces [of Areas] procedure discusses *xu* and *yi*. Moreover, there is no specific vocabulary for signs in the Old Procedure. It is possible to discern several layers of composition within the procedure. The ancient authors of the tenth and eleventh centuries, Liu Yi and Jiang Zhou, dealt with negative numbers in relation to 'empty' and 'augmented' areas. When Li Ye and Yang Hui collated these sources and composed their own texts, they already had different ideas about these concepts. The various aspects of the procedure described here demonstrate the elaboration of the concept of equation with negative coefficients. Their treatises, in fact, testify to meandering developments in the construction of mathematical concepts and their reception by later readers.

Conclusion to Part II

The study of the transformation of figures calls into question the concept of negative and positive coefficients. Through the imaginary movement of areas, the movement of emptying or filling areas, and the movement of cutting or assembling rectangles into ribbon, a progression of expressing equations emerges. By Liu Yi's time, distinct expressions referred to negative and positive quantities. However, because equations are based on the construction of diagrams and negative areas are not intuitively obvious, the geometric elaboration of negative coefficients presented interesting issues. The positive and negative quantities are treated independently on the counting surface; the equation represents a perceived correspondence between tabular exposition and diagrammatic representation. However, the distinctive elements are not clearly stated: negative coefficients, equalities, and unknowns are all absent from the discourse. The evolution of computational practices with diagrams, especially the geometric representation of a mathematical object called 'adjunct' which was filled or emptied, ensured that the negative quantities became associated with negative coefficients. At the time of Li Ye and Yang Hui, negative coefficients were clearly constructed and quadratic equations were free from geometrical representations. In their treatises, Li Ye and Yang Hui present several vignettes in the elaboration of quadratic equations with negative coefficients.

A comparison of the procedures in the two treatises displays a process of maturation at play in the mathematical concepts. Such a comparison also shows that several procedures with the same name in different texts can mask divergent practices. Hence, it is important to distinguish the elaboration of mathematical concepts (in this case, from the eleventh century) from their transmission and reinterpretation (from the thirteenth century).

The procedure attributed to Liu Yi by Yang Hui presents more elaborate mathematical objects than the Old Procedure presented by Li Ye in the *Development of Pieces [of Areas]*. The Old Procedure presents a specific pattern of a single adjunct in a diagram. The various types of procedure show a change in the use of the diagrams, from complex drawings to simpler compositions with more direct visualisations of the figures. The diagram disappeared first from the root extraction, and subsequently from the elaboration of the equation. The pattern of diagrams of a single adjunct appears in other problems of *Yang Hui's Methods of Computation*, but totally disappears from the Section of Pieces [of Areas] procedure of the *Development of Pieces [of Areas]*. This suggests that the Old Procedure is older than the procedure preserved in *Yang Hui's Methods of Computation*, and that the procedure of *Yang Hui's Methods of Computation* is older than that of the Section of Pieces [of Areas]. Consequently, there should be a reconsideration of some chronological elements from the introduction. Li Ye added a 'recommendation' instead of changing the 'original' procedure. The vocabulary used in his commentary differs from that of the mathematical discourse. Taken together, these facts indicate that there are several layers of composition to the *Development of Pieces [of Areas]*, which means that more than the twenty-three Old Procedures remaining in Jiang Zhou's *Collection Augmenting the Ancient [knowledge]* were preserved in the *Development of Pieces [of Areas]*. Nonetheless, the origin of the Old Procedure remains vague: was it a reconstruction by Li Ye or an ancient artifact already preserved in the *Collection Augmenting the Ancient [knowledge]*? Why, then, is Li Ye writing a 'recommendation' instead of directly correcting the procedure?

Problem 2, 18, 5

Problem 2

第二問

[2.1] 今有方田一段, 内有圓池. 水占之計外[17] 地一十三畝七分半. 竝不記徑面. 只云從外田南楞通内池北楞四十步.
問内圓外方各多少.
荅曰: 同前.

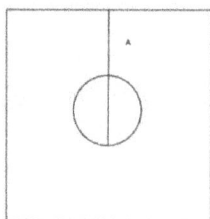

Fig. 6.5　a: 通池徑四十步

[2.2] 法曰: 立天元一[18] 為池徑. 減倍通步得 ⊞. 為田方面.

[2.3] 以自增乘得 ⊞ 為方田積, 於頭.

[2.4] 又以天元池徑. 自之, 三因, 四而一得 ⊞ 為池積.

[2.5] 以減頭位得 ⊞ 為一段虛積, 寄左.

[2.6] 然後列真積三千三百步. 與左相消得者 ⊞. 開平方得二十步, 即内池徑也. 倍通步内減池徑為方面也.
[2.7] 依條段求之. 倍通步自乘於頭位. 以田積減頭位, 餘為實. 四之通步為從. 二分半虛常法.

---
[17]外計 in WYG.
[18]— is not in WYG and WJG

124 The Empty and the Full: Li Ye and The Way of Mathematics

[2.8] 義曰: 倍通步者, 是於方面之外引出一圓也. 用二分半虛常法者, 是一個虛方內卻有減餘圓池. 補了七分半, 外欠二分半. 故以之為虛隅也.

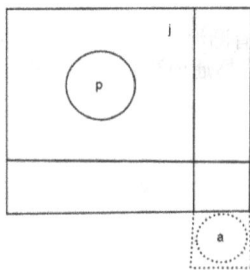

Fig. 6.6   j: 減; a: 七分五釐; p: 池.

Translation:

Problem 2.

[2.1] Let us suppose there is one piece of square field, inside which there is a circular pond. Outside the [area] occupied by water, one counts thirteen *mu* and seven *fen* and a half of land. Moreover there is no record of the diameter.[19] It is said only that [the distance] from the south edge of the outer field through the north edge of the inside pond is forty *bu*.

One asks how much are the diameter of inner circle and the side of the outer square each.

The answer: same as before [i.e. same as Problem 1].

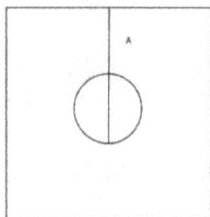

Fig. 6.7   a: through the diameter of the pond, forty *bu*.

---

[19]經面 *jing mian*. The standard expression for 'diameter' is 經.

[2.2]The method: Set up one[20] Celestial Source as the diameter of the pond. Subtracting it from twice the *bu through* yields $\begin{smallmatrix} 8 & 0 & tai \\ & -\,1 & \end{smallmatrix}$ as the side of the square field.

[2.3] Augmenting this by self-multiplying yields $\begin{smallmatrix} 6\,4\,0\,0\ tai \\ -\,1\,6\,0 \\ 1 \end{smallmatrix}$ as the area of the square field, which is sent to the top position.

[2.4] [Set up] further the Celestial Source, the diameter of the pond. This times itself, then increased by three, and divided by four yields $\begin{smallmatrix} 0 & tai \\ 0 & \\ 0.7\,5 & \end{smallmatrix}$ as the area of the pond.

[2.5] Subtracting this from the top position yields $\begin{smallmatrix} 6\,4\,0\,0 & tai \\ -\,1\,6\,0 & \\ 0.2\,5 & \end{smallmatrix}$ as one section of empty area, which is sent to the left.

[2.6] Afterwards, one places the real area of three thousand and three hundred *bu*. With what is on the left, eliminating from one another yields $\begin{smallmatrix} -3\,1 & 0\,0 \\ 1 & 6\,0 \\ -\,0.2\,5 & \end{smallmatrix}$.

Opening the square yields twenty *bu*. That is the diameter of the inside pond. From twice the *bu through*, one subtracts the diameter of the pond to make the side of the square.

[2.7] One looks for this according to the section of pieces [of areas]. One self-multiplies twice the *bu through* and places this on the top position. One subtracts the area of the field from what is on the top position. The remainder makes the dividend. Four times the *bu through* makes the adjunct. Two *fen* and a half is the empty constant divisor.

[2.8] The meaning: Twice the *bu through*, that is to extend, outside the side of the square, one diameter of the circle.

[2.9] To use an empty constant divisor of two *fen* and a half, that is, conversely, inside of one empty square, to have a subtraction of the circular pond that remained. After compensating with seven *fen* and a half, at the outside, it lacks two *fen* and a half. Therefore, with this, one makes the empty corner.

Description.

[2.1] Let a be the distance from the middle of the side of the square going through the pond, 40 *bu*; let A be the area of the square field (S) less the area of the circular pond (C), 3300 *bu*; and x be the diameter of the pond.

---

[20]The character 一 is not in the WYG and WJG *siku quanshu*: 'set up a Celestial Source'

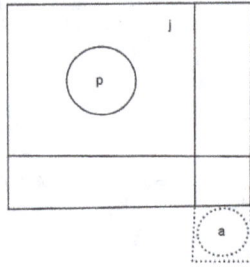

Fig. 6.8   j: to subtract; a: seven *fen* five *li*; p:.pond.

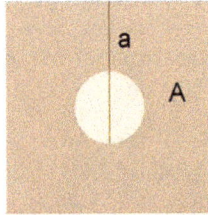

Fig. 6.9

The procedure of the Celestial Source:

[2.2] Side of the square $= 2a - x = 80 - x$.

[2.3] $S = (2a - x)^2 = 4a^2 - 4ax + x^2 = 6400 - 160x + x^2$.

[2.4] $C = \frac{3}{4}x^2 = 0.75x^2$, since $\pi = 3$.

[2.5] $S - C = 4a^2 - 4ax + x^2 - \frac{3}{4}x^2 = A = 6400{-}160x + x^2 - 0.75x^2 = 6400{-}160x + 0.25x^2 = 3300$ *bu.*

[2.6] The equation: $A - (4a^2 - 4ax + 0.25x^2) = -3100 + 160x - 0.25x^2 = 0$.

The procedure by Section of Pieces of Areas:

According to the description of terms given by the discourse, $4a^2 - A$ is the dividend; $4ax$ is the adjunct and $-0.25x^2$ is the constant divisor.

The reader constructs a square with twice the known distance a, which corresponds to $4a^2$, and from this removes the known area A to make the constant term. [Fig. 6.10] represents $4a^2 - A$, the known area, or constant term of the equation. The area that

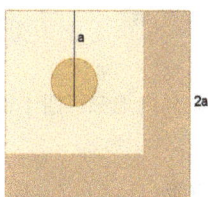

Fig. 6.10

Fig. 6.11

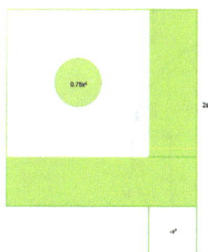

Fig. 6.12

Fig. 6.13

remains can also be interpreted as two rectangles of length $2a$ and width x: these two rectangles represent the adjunct divisor, $4ax$, and have in common one square with the side equal to the diameter of the pond. [Fig. 6.11] represents the two rectangles which are stacked on one square. Once one has 'un-stacked' these two areas, one obtains the [Fig. 6.12]. Li Ye explains this by [2.8] '*Twice the bu through, that is to extend outside the side of the square one diameter of the circle*'. This square of side x is supplementary, thus one removes it. Afterwards, as in Problem 1, the circular pond has to be removed from that square. Since the square is already removed, that is to say 'empty', to remove one circular pond amounts to adding it instead. That is: $-x^2 + 0.75x^2$ [Fig. 6.13]. This operation makes that the remaining space between the circle and that square equals to $-0.25x^2$. Li Ye names this '*to compensate*'; and the result of the operation 'is lacking'. Li Ye explains this in the following way [2.9]: '*To use an empty constant divisor of two fen and a half, that is, conversely, inside of one empty square, to have a subtraction of the circular pond that remained. After compensating with seven fen and a half, at the outside it lacks two fen and a half, which, therefore, makes the empty corner*'. So we read the equation as $4a^2 - A = 4ax - 0.25x^2$.

Problem 18

第十八問

[18.1] 今有圓田一段, 內有方池. 水占之外計地三百四十七步. 只云外圓周內方周共得二百八步.

問內外周各多少.

荅曰: 外圓周一百八步. 內方周一百步.

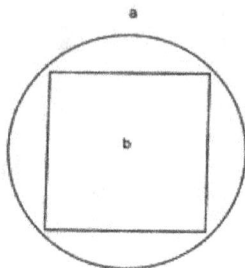

Fig. 6.14   a: 圓田; b: 二十五步.

[18.2] 法曰: 立天元一為內方面. 以四之為內方周.

[18.3] 減於相和二百八步得 ⊞ 為外圓周.

[18.4] 以自增乘得 ⊞ 為圓周冪, 便為十二段圓田積, 於頭.

[18.5] 再立天元內方面. 以自之, 又就分十二之得 ⊞ 為十二段方池積也.

[18.6] 以減頭位餘 ⊞ 為十二段如積, 寄左.

[18.7] 然後列見積三百四十七步. 就分母十二之得四千一百六十四步.

[18.8] 與左相消得 ⊞ .

[18.9] 開平方得二十五步為內方面也. 四之為內方周. 減於相和步為圓周也.

[18.10] 依條段求之. 以十二之積步, 減和步冪為實. 八之和步為虛從. 四常法.

[18.11] 義曰: 十二段圓田內有十二個方池. 於方周冪內, 補了十二池. 外

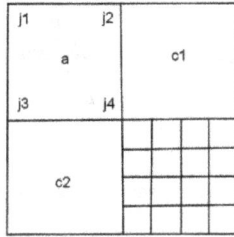

Fig. 6.15　a: 此外圓周冪也該十二圓田積; c1: 連下十六池面為四之和步從; c2: 連右十六池面為四之和步從; j1–4: 減.

猶欠四個. 故以四為隅法.

[18.12] 此式元係虛從. 今卻為虛隅, 命之故以四為虛常法.

[18.13] 舊術曰: 相和步自乘. 於頭位. 以十二之積步. 減頭位餘八. 而一為實. 相和步為從法. 廉常置半步. 減從.

Translation:

Problem 18.

[18.1] Let us suppose that there is one piece of a circular field inside which there is a square pond. Outside the [area] occupied by water, one counts three hundred forty-seven *bu* of land. It is said only that the circumference of the outer circle and the perimeter of the inside square [added] together yields two hundred eight *bu*.

One asks how much are the outer circumference and inner perimeter each.

The answer: the circumference of the outer circle is one hundred eight *bu* and the perimeter of the inside square is one hundred *bu*.

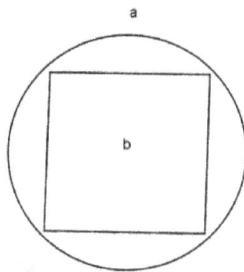

Fig. 6.16　a: circular field; b: Twenty five *bu*.

[18.2] The method: Set up one *Celestial Source* as the side of the inside square. This (multiplied) by four makes the perimeter of the inside square.

[18.3] Subtracting this from the *mutual sum* (*xiang he* 相和) of two hundred eight *bu* yields $\begin{smallmatrix} 2 & 0 & 8 \ tai \\ & -4 \end{smallmatrix}$ as the circumference of the outer circle.

[18.4] Self-multiplying this by increasing yields $\begin{smallmatrix} 4 \ 3 \ 2 \ 6 \ 4 \\ -1 \ 6 \ 6 \ 4 \\ 1 \ 6 \end{smallmatrix}$ as the square of the circumference of the circle, which then makes twelve pieces of the area of the circular field, and which is sent to the top.

[18.5] Set up again the *Celestial Source*, the side of the inside square. Self-multiply this and, with the help of parts, this by twelve yields $\begin{smallmatrix} 0 \ yuan \\ 1 \ 2 \end{smallmatrix}$ as the twelve pieces of the area of the square pond.

[18.6] Subtracting from what is on the top position, $\begin{smallmatrix} 4 \ 3 \ 2 \ 6 \ 4 \\ -1 \ 6 \ 6 \ 4 \\ 4 \end{smallmatrix}$ remains as the twelve pieces of the equal area, which are sent to the left.

[18.7] Afterwards, place the real area: three hundred forty-seven *bu*. With the help of parts, [multiplying] this by the denominator twelve makes four thousand, one hundred sixty-four *bu*.

[18.8] With what is on the left, eliminating them from one another yields $\begin{smallmatrix} 3 \ 9 \ 1 \ 0 \ 0 \\ -1 \ 6 \ 6 \ 4 \\ 4 \end{smallmatrix}$

[18.9] Opening the square yields twenty five *bu* as the side of the inside square. This [multiplied] by four makes the perimeter of the inside square. Subtracting it from the *bu* of the *mutual sum* makes the circumference of the circle.

[18.10] One looks for this according to the Section of Pieces [of Areas]. [From] the square of the *bu* of the *sum*, one subtracts twelve times the *bu* of the area to make the dividend. Eight times the *bu* of the *mutual sum* makes the empty adjunct. Four is the constant divisor.

[18.11] The meaning: Inside the twelve pieces of the circular field, there are twelve square ponds. Inside the square of the perimeter of the square, once the twelve ponds are compensated, outside, in what remains, it lacks four [ponds]. Therefore, with four, one makes the corner divisor.

[18.12] The configuration originally empties the adjunct. Now, on the contrary, the corner is emptied. That is why I recommend that four makes the empty constant divisor.

[18.13] The Old Procedure: self-multiply the *bu* of the mutual sum; place them in the top position. [Multiply] the *bu* of the area by twelve; subtract them from what is in the top position. Divide the

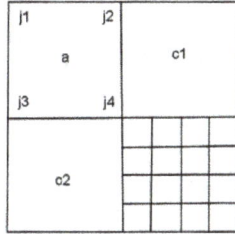

Fig. 6.17 a: 'This is the square of the circumference of the outer circle; it produces twelve areas of the circular field.' c1:, 'Below are sixteen ponds; the side of the square makes four times the *bu* of the mutual sum. That is the adjunct.' c2: 'On the right are sixteen ponds; the side of the square makes four times the *bu* of the mutual sum. That is the adjunct.' j1–4: 'Subtract'.

remainder by eight to make the dividend. The *bu* of the mutual sum makes the adjunct. There is a divisor and an edge. The constant [divisor] is half a *bu*. Subtract the adjunct.

Description.

[18.1] Let a be the sum of the circumference and the perimeter, 208 *bu* ('the *bu* of the sum' or 'the *bu* of the mutual sum'); let A be the area of the circular field (C) minus the area of the square pond (S), 347 *bu* ('the *bu* of the area'); and let x be the side of the pond.

Fig. 6.18

The Procedure of the Celestial Source:

[18.2] Perimeter of the square $= 4x$.
[18.3] Circumference $= a - 4x = 208 - 4x$.

[18.4] Square of the circumference $= (a - 4x)^2 = a^2 - 8ax + 4x^2 = 43264 - 1664x + 16x^2 = 12C$.

[18.5] $12S = 12x^2$.

[18.6] Because $12C - 12S = 12A$, $12C - 12S = a^2 - 8ax + 4x^2 - 12x^2 = 43264 - 1664x + 16x^2 - 12x^2 = 43264 - 1664x + 4x^2$.

[18.7] $12A = 4164$ *bu*.

[18.8] The equation: $a^2 - 12A - 8ax + 4x^2 = 0$. With the values of the coefficients: $39100 - 1664x + 4x^2 = 0$.

[18.9] The side of the square: $x = 25$.
The perimeter: $4x = 100$. The circumference: $a - 4x = 208 - 100 = 108$.

Problem 5

第五問.

[5.1] 今有方田一段, 內有圓池水占, 之外計地一十三畝二分. 只云內圓周不及外方周一百六十八步. 問方圓各多少. 荅曰: 外方周二百四十步. 內圓周七十二步.

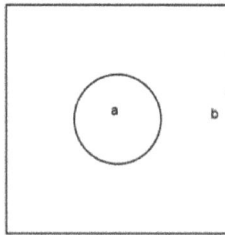

Fig. 6.19   a: 池徑, 二十四步; b: 田方, 六十步.

[5.2] 法曰: 立天元一為內圓周. 加一百六十八步得 ⊔ 為外方周.

[5.3] 以自增乘得 ⊔ 為一十六個方田積.

[5.4] 又三因之得 ⊔ 為四十八段方田積, 於頭. [所以三因之為四十八者就為四十八分母也.][21]

---

[21]Sentences written in smaller, bracketed characters are commentaries attributed

[5.5] 再立天元圓周. 以自之 █ 為十二段圓池積. [圓周羃為九個圓徑羃. 每三個圓徑羃為四個圓池積. 今九個圓徑羃共為十二個圓池積也.]

[5.6] 又就分四之得 █ 為四十八個圓池積.

[5.7] 以減頭位得 █ 為四十八段如積, 寄左.

[5.8] 然後列真積一十三畝二分. 以畝法通之得三千一百六十八步. 又就分母四十八之得一十五萬二千零六十四步.

[5.9] 與寄左相消得 █. 平方開之得七十二步為內圓周也三而一為池徑.

[5.10] 依條段求之. 四十八段田積內減三段不及步羃為實. 六之不及為從. 一虛隅.

[5.11] 義曰: 每一個方周方為十六段方田積. 今三之為四十八段方田積也. 內除了三個圓周羃. 外於見積上, 虛了一個圓周羃也. 今求圓周, 故以一步為虛隅法.

[5.12] 舊術曰: 以十六乘田積為頭位. [以合方周之積] 以不及步自乘減頭位, 餘三之為實. 六之不及步為從法, 廉常以一步為減從法.

Translation:

Problem 5.

[5.1] Let us suppose that there is one square piece of field, inside which there is a circular pond. Outside the [area] occupied by water, one counts thirteen *mu* two *fen* of land. It is said only that the circumference of the inner circle does not attain the perimeter of the outer square by one hundred sixty-eight *bu*.
One asks how much are the circumference and the perimeter each. The answer: the perimeter of the outer square is two-hundred-forty *bu*; the circumference of the inside circle is seventy-two *bu*.

[5.2] The method: Set up one *Celestial Source* as the circumference of the inside circle. Adding one-hundred-sixty-eight *bu* yields
1 6 8 *tai*
  1   as the perimeter of the outer square.

[5.3] Augmenting this by self-multiplying yields
2 8 2 2 4 *tai*
  3 3 6   as sixteen
    1
areas of the square field.

Fig. 6.20   c1–6: 從; j1–3: 滅.

Fig. 6.21   a: Diameter of the Pond: Twenty-Four *bu*; b: Side of the Field: Sixty *bu*.

$$8\ 4\ 6\ 7\ 2\ tai$$

[5.4] Multiplying further by three yields $\dfrac{1\ 0\ 0\ 8}{3}$ as forty-eight

pieces of area of the square field, which is sent to the top.

The reason of the multiplication by three and of making of forty eight is that it makes forty eight for denominator.[22]

---

[22]Paragraph in italic are commentaries written in smaller characters in the Chinese

[5.5] Set up again the *Celestial Source* as the circumference of the circle. Self-multiplying this yields $\begin{smallmatrix} 0 \\ 1 \end{smallmatrix}$ *yuan* as twelve pieces of area of the circular pond.

The square of the circumference of the circle makes nine squares of the diameter of the circle. Each three squares of the diameter of the circle make four areas of the circular pond. Now, nine squares of the diameter of the circle together make twelve areas of the circular pond.

[5.6] With the help of the parts, quadruple this. It yields $\begin{smallmatrix} 0 \\ 4 \end{smallmatrix}$ *yuan* as forty-eight areas of the circular pond.

[5.7] Subtracting this from what is on the top position yields

8 4 6 7 2 *tai*
1 0 0 8   as forty-eight pieces of the equal area, which is sent
− 1

to the left.

[5.8] Afterwards, place the real area of thirteen *mu* two *fen*. With the divisor of *mu*, making this communicate, it yields three-thousand-one-hundred-and-sixty-eight *bu*. With the help of the denominator, one multiplies further by forty-eight. It yields one-hundred-fifty-two-thousand-and-sixty-four *bu*.

[5.9] With what is on the left, eliminating from one another yields

−6 7 3 9 2
1 0 0 8. Opening the square of this yields seventy-two *bu* as
− 1

the circumference of the inside circle. Dividing by three makes the diameter of the pond.[23]

[5.10] One looks for this according to the Section of Pieces [of Area]. From the forty-eight pieces of area of the field, subtract three pieces of the square of the *bu that does not attain* to make the dividend. Six times the *[bu] that does not attain* makes adjunct. The empty corner is one.

[5.11] The meaning: Each of the squares of the perimeter makes sixteen pieces of area of the square field. Now, tripling [it] makes forty-eight pieces of area of the square field. Inside, once the three squares of the circumference of the circle are removed from the area that appears outside, one square of the circumference of the circle is emptied.

Now, one looks for the circumference of the circle; that is why one takes one *bu* to make the empty corner-divisor.

[5.12] The Old Procedure: the area of the field multiplied by sixteen makes what is in the top position.

edition.
[23] Only the diameter was given. The perimeter and the circumference were to be found.

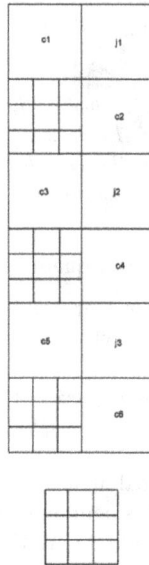

Fig. 6.22   c1–6: adjunct. j1–3: Subtract.

It corresponds to the area [made by the square] of the perimeter.
Multiply by itself the bu that does not attain, subtract them from
what is on the top position, and triple what remains to make the
dividend. Six times the bu that does not attain makes the adjunct.
There is a divisor and an edge. One bu is taken to make the con-
stant [divisor]. Subtract the adjunct divisor.

Description:

[5.1] Let a be the difference between the perimeter (p) of the square
and the circumference (c) of the pond, 168 *bu*. Let A be the area
of the square field (S) less the area of the circular pond (C), 13 *mu*
2 *fen*, or 3168 *bu* (1 *mu* = 240 *bu*). Let x be the circumference.
The Procedure of the Celestial Source:

[5.2] $p = a + x = 168 + x$.
[5.3] $p^2 = (a + x)^2 = a^2 + 2ax + x^2 = 28224 + 336x + x^2 = 16S$.
[5.4] $3p^2 = 3 \times 16S = 3a^2 + 6ax + 3x^2 = 84672 + 1008x + 3x^2 = 48S$.
[5.5] $12C = x^2$. In commentary: Let c be the circumference; d, the
diameter; and C, the area of the pond. $c^2 = x^2 = 9(d^2)$; $3(d^2) = 4C$; $9(d^2) = 12C$.
[5.6] $12C \times 4 = 4x^2 = 48C$.

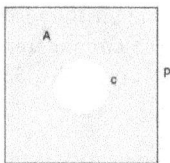

Fig. 6.23

[5.7] $48S - 48C = 3a^2 + 6ax + 3x^2 - 4x^2 = 3a^2 + 6ax - x^2 = 48A$.

[5.8] $84672 + 1008x - x^2 = 152064$ *bu*.

[5.9] The equation: $3a^2 - 48A + 6xa - x^2 = 0$ or, $-67392 + 1008x - x^2 = 0$.

The Procedure by Section of Pieces [of Area]:

According to [5.10], 'from forty-eight pieces of area of the field, subtract three pieces of the square of the *bu that does not attain* to make the dividend. Six times the *[bu] that does not attain* makes the adjunct. The empty corner is one,' as reported in [Table 6.5].

Table 6.5

$48A - 3a^2$	dividend
$6a$	adjunct
1	empty corner

Following the instructions in [5.11] and in the diagram, the following procedure emerges: Three squares for which the sides are the perimeter (p) are constructed [Fig. 6.24]. Each of these squares is $16S$, therefore three squares are $48S$. Li Ye writes that 'each square of perimeter makes sixteen pieces of area of the square field. Now, to multiply this by three makes forty-eight pieces of area of the square field'. Given that $48S = 48C + 48A$, and that $48A$ is given in the statement, to find $48A$, remove $48C$ from $48S$. Also, $3(p^2)$ are three squares of side $a + x$.

This area is composed of six adjunct rectangles of length x and width a, and three squares of side a. These rectangles are superimposed on a square area of side x. The area may be expressed as: $48S = 3a^2 + 6ax + 3x^2$. [Fig. 6.25]

To remove $48C$, the fact that $c^2 = 9(d^2) = 12C = x^2$ can be used. Therefore, each square of side x is twelve ponds (Li Ye represented by nine little squares of the diameter in his diagram). First, three squares are removed 'from the visible area'. This operation removes 36 ponds. To remove the twelve other ponds is equivalent to removing a square of side c at the '*outside*'. [Fig. 6.26] Li

Ye writes that, 'inside, once one has removed three squares of the circumference of the circle on the visible area, at the outside, one empties one square of the circumference of the circle'. The result is: $48A = 3a^2 + 6ax - x^2$; or $48A - 3a^2 = 6ax - x^2$.

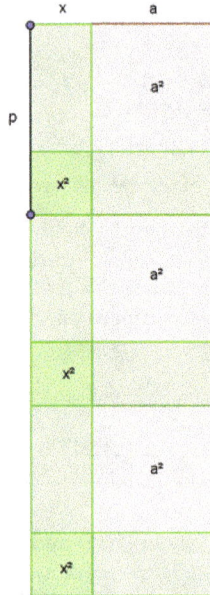
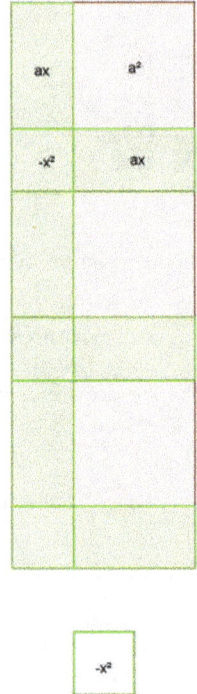

Fig. 6.24   Fig. 6.25   Fig. 6.26

Part III: Order, Analogy and Reduction

The order in which the problems of *Development of Pieces [of Areas]* are presented also contains clues to the development of mathematical concepts. The sixty-four problems are not randomly ordered, nor do they reflect increasing complexity. The construction of the diagram helps to reveal the logic behind the order in which the problems are presented.

In fact, the reasons for the arrangement of the problems are precisely what give a meaning to the treatise. Several modes of organisation combine together to construct the sophisticated architecture which directs the order of the problems. After solving each problem by means of geometric visualisation, the practitioner reaches a clear understanding of the validity of the geometrical procedure and its generality. The strengths and limits of the procedure are demonstrated through silence, whereby seeing is enough to grasp why something is correct.

The impossible challenge of this chapter is to describe in words a practice which was intended to be communicated non-verbally! Here, proofs build upon movements, while words serve to establish static relationships. I hope the reader will benefit from executing each of the sixty-four problems, as Li Ye required, because experience constitutes the only way to reach a shared signification.

The first chapter describes the visibly manifest part of the classification of problems. The primary organisation relies on shapes (circles, squares···) as well as the construction of distances given in the statement of problems, in which some data were reduced to make problems similar. The question of similarity forms the object of the second chapter, which uses analogy to guide the practitioner through the solution for each problem. This analogy requires a specific rhetorical approach to solve the problems. The last chapter unveils the general structure of the text (as far as possible) and explains how the visualisation of structures was used as an expression of generality.

Chapter 7

Statement of the Problems: The Example of Problem 3

The key to understanding the content of *Development of Pieces [of Areas]* is not its narrative thread. The structure of the treatise reveals its most important concern. The problems of the *Development of Pieces [of Areas]* are not randomly ordered. There is a system of classification based on a field survey of problem types which is a pretext for abstract thought. In some of the problems, the numerical data are transformed into other types of data. The purpose is to simplify the problem and thereby link the problem with previously solved problem. This method consists of reducing the data of a sophisticated problem to a previously worked simple example. This method reveals a primary classification of problems according to the data given in their statements. This chapter describes the process of reduction.

The purpose of the *Development of Pieces [of Areas]* is conveyed in the order of its problems. Solving mathematical problem is a pretext for something else. For each of the problems, the statement appears to be a problem of land surveying. There is a field containing a pond; only the shapes change from one problem to another. The statement is always enunciated in the same way: the problems are numbered, the shapes of the field and pond are given, with the difference between their areas and a distance being given. A diagram accompanies the statement, which ends with a question and the immediate answers to the problems, as seen in problems provided as examples in previous chapters. This is a classical way of writing problems in ancient Chinese-language mathematical texts [Chemla (2000)]. This tradition of writing leads one to question the status of problems: if the purpose of mathematical books from China in general and the *Development of Pieces [of Areas]* in particular is problem solving, why is the solution given immediately? What is the status of a problem for which the answer is instantly available? It seems clear that the answer to the problem is not the main purpose for the inclusion of the problem. Understanding the practice of ordering problems will answer this question and lead to a revision of the status of the *Development of Pieces [of Areas]*. This first investigation addresses the order directly governed by the data of statement, while the next analysis concentrates on the order implied by the solutions of problems.

7.1   Order of Data in the Statement of Problem

At first glance, the problems are ordered according the geometric shape proposed in the statement. Although this organisation holds for the first twenty problems, the organisation of the remainder of the book relies on obscure principles. [Table 7.1] lists of the types of geometric figures following the order of problems:

Table 7.1: The geometrical shapes in the statement of a problem

Chapter 1	Problems 1–10	A circular pond in the centre of a square field
	Problems 11–20	A square pond in the centre of a circular field
	Problem 21	Three square fields of different size
	Problem 22	A square field with a triangular pond in one corner
Chapter 2	Problems 23–29	A square and a circular field next to each other
	Problem 30	Two circular fields
	Problem 31	A rectangular field with a circular pond in the centre
	Problems 32–37	A circular field with a rectangular pond in the centre
	Problem 38	Two rectangular fields next to each other
	Problems 39–42	A rectangular field with a circular pond in the centre
Chapter 3	Problem 43	Three circular fields of different sizes with different value of π

	Problem 44	A trapezoid field
	Problem 45	A square field with a square pond in the centre
	Problem 46	A square and a circular field next to each other
	Problem 47	A rectangular field with a square pond in the centre
	Problem 48	A square field with a rectangular pond in the centre
	Problems 49–52	A square field with a square pond in the centre
	Problems 53–54	A square field with a rectangular pond in the centre
	Problems 55–56	A circular field with a circular pond in the centre
	Problems 57–58	A circular field with a rectangular pond in the centre
	Problem 59	A square field with a circular pond in the centre which has a square field in its centre
	Problem 60	A circular field with a square pond in the centre which has a circular field in its centre
	Problem 61	A square field with a circular pond at one corner
	Problem 62	A square field with a square pond at one corner
	Problem 63	A big circular field, a big square field and a small square field with a circular pond in its centre
	Problem 64	A square field with a concentric pond in the centre

At first sight, problems present statements and solutions independently of each other. While Chapter 2 and Chapter 3 present quite different problems, the problems are ordered according to the statement of their geometrical shapes. This is clear in Chapter 1: Problems 1 to 10 deal with circular ponds inside square fields whereas Problems 11 to 20 present square ponds inside circular fields. In Chapter 3, although Problem 59 presents a square field with a circular pond in the centre, that in turn has a square field in its centre, the following problem proposes an inverse figure: a circular field with a square pond in the centre that in turn has a circular field in its centre. All problems treating the same kind of geometric shape are grouped together. The statements of problems are regrouped in categories constructed according to shape and then set in order. However, a closer reading of the *Development of Pieces [of Areas]*, reveals the order of the problems is less clear. Chapter 3 contains a haphazard list of all possible figures, like variations on the topic of a field and a pond. Here, another sub-classification may direct the arrangement of the problems.

Within the categorisation of shapes of figures, another classification of problems can be made to explain the variations. Each statement of the problem is composed of two items of data: an area and a distance. The statement first situates the two figures with respect to each other and then gives the area resulting from the difference between the areas of the two figures.[1] Afterwards, one distance is given as the result of several operations on two segments, one from the field and one from the pond, respectively. The two other segments form the solution to the problem and are given as answers. This distance has been termed 'a' in the mathematical transcriptions given in examples.

For example, Problem 49 reads as follows:

Problem 49.

> [49.1] Let us suppose there is one square piece of field, inside which there is a small square pond settled in lozenge.[2] Outside the [area] occupied [by water], one counts ten thousand eight hundred *bu* of land. It is said only [that the distances] from the edge of the outer field reaching the angle of the inner pond are eighteen *bu* each. One asks how much are the sides of the outer and the inner squares. The answer: the side of the outer square field is one hundred twenty *bu*. The side of the inner square pond is sixty *bu*. [3]

In Problem 49, the given area is the result of the subtraction of the square lozenge from the area of a bigger square. The given segment is half

---

[1] For few of the problems, there are three figures. In those cases, either the sum of their areas, or the sum of two areas minus the third one, is given.

[2] 結角, jie jiao.

[3] 第四十九問

今有方田一段, 內有小方池結角. 占之外計地一萬八百步. 只云從外田楞至內池角各一十八步. 問內外方各多少.

荅曰: 外田方一百二十步. 內池方六十步.

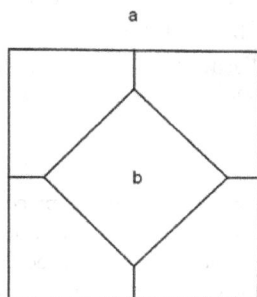

Fig. 7.1   a: square field; b: square pond.

of the side of the outer square diminished by the diagonal of the square lozenge. If one tries to express the distance given in the statement according to the two other distances (here, a side and a diagonal) required by the problem, it can be transcribed as $(side - diagonal)/2$. This transcription is a bit different from the way in which Li Ye states the data, but it reveals a clue to the patterns of classification. A list of the transcription of the distances given in each statement appears in [Table A.2 in Appendix A].

In [Table A.2], the same sequence of operations used to construct the distance given in the statement is used recurrently. That is, one segment of the outer field has one segment of the inner pond subtracted from it. Then, the same segment of the outer field has the same segment of the inner pond added to it, with some variations on this pattern. Whatever the geometrical shapes of the field and pond, the statements of the problem are presented as an iterative list of possible constructions from this distance, and the same construction sequence is observed for each of the geometric categories. As for the order of problems according to their statement, first there is a regrouping of problems according their geometrical shape, that is a grouping according to areas. Secondly, the groups of problems are ordered according the construction of the second item of data, the distance.

7.2   Reduction of Data and Procedure

The construction of the distance 'a' reveals another aspect of this classification of objects. Besides the listing of objects, there is a process of reduction of data in order to relate solutions to one another. For the majority of problems, the two constants (area and distance) are immediately given and used to solve the problem. The reader operates directly with these data and the chosen unknown. The distances sought in answer are also immediately generated. For example, in the following case of Problem 3:

Problem 3.[4]

> [3.1] Let us suppose there is one square piece of field, inside which there is a circular pond. Outside [the pond] occupied by water, one counts eleven thousand three hundred twenty-eight *bu* of land. It is said only that [the distances] from the angle of the outer square obliquely reaching the edge of the inner pond are fifty-two *bu* each. One asks how much are of the inside diameter and the outer side each.
> The answer: the side of the outer field is one hundred twenty *bu*. The diameter of the inside pond is sixty-four *bu*.

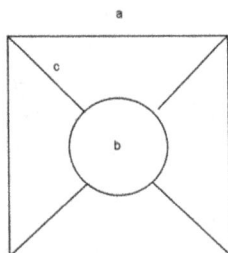

Fig. 7.2   a: square field; b: circular pond; c: fifty-two *bu*.

> [3.2] The method: Set up one *Celestial Source* as the diameter of the inside pond. Adding twice the reaching *bu* yields $\begin{smallmatrix} 1\ 0\ 4\ tai \\ 1 \end{smallmatrix}$ as a diagonal of the square.

In Problem 3, the data given in the statement in [3.1] are directly used to construct the first polynomial. 'The method' starts with the distance given in the statement multiplied by two and added to the diameter, which is later demanded. This direct use of data is not required in each of the problems, though. Two exceptions are interesting from this point of view.

In two cases (Problems 19 and 64), the data are transformed to derive other quantities which are preferred by Li Ye or which were first 'non-usable' and then transformed into 'usable' form.

---

[4]Chinese text and complete translation are given at the end of this chapter.

For example, in Problem 19, the data in the statements are an area resulting from the difference between an outer circular field and an inner square pond and the sum of the 3 segments. Li Ye added a commentary to the Celestial Source procedure in [19.3]:

Problem 19.

[19.1] Let us suppose there is one circular piece of field inside which there is a square pond. Outside the [area] occupied by water, one counts thirty-three *mu*, one hundred seventy-six *bu* of land. It is said only that the outer circumference, the inner perimeter and [the distance that] crosses the area mutually summed up together yields six hundred two *bu*.

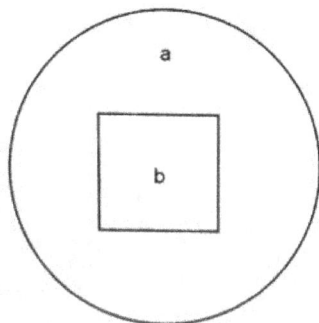

Fig. 7.3 a: the diameter crossing [the area] is thirty four *bu*. 實徑三十四步; b: pond. 池.

[19.2] The method: Set up one *Celestial Source* as the side of the inside square. Subtracting it from one hundred seventy-two yields $\begin{smallmatrix} 1 \ 7 \ 2 \ tai \\ -\ 1 \end{smallmatrix}$ as the diameter of the outer field.

[19.3] Twice the quantity that is mentioned yields one thousand two hundred four *bu*, which is in other words: six diameters of the circle, eight sides of the square, and two [distances that] cross the area. Now, if one sets up one side of the square and two [distances that] cross the area; their sum becomes one diameter. The quantity counted above is seven sides of the square and seven diameters of the circle. Now, one places one thousand two hundred four *bu* at the earth [position]. Reducing this by seven yields one hundred seventy-two *bu* as the sum of the side and the diameter, which becomes one side of the square and one diameter of the circle, with no [distances which] crosses the area.' [5]

Here, the statement and the commentary are transcribed in modern terms:

[19.1] Let c be the sum of the circumference, the perimeter and the distance from the middle of the side of the square to the circle, i.e. 602 *bu*; let A be the area of the circular field (C) minus the area of the square pond (S), i.e. 33 *mu* 176 *fen* and let x be the side of the square pond.

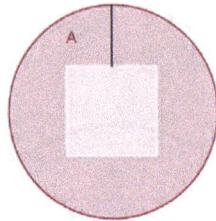

Fig. 7.4

[19.3] explains the following: c is the sum of circumference, the perimeter, and the distance from the square to the circle. Let d be the diameter, s be the side and b be the distance from the square to the circle. Then $2c = 6d + 8s + 2b = 1204$. It is known that $1d = s + 2b$. This means that $2c = 7s + 7d$, or $2c/7 = s + d = 172$. This quantity may be named a. This quantity a will be used for constructing the polynomials corresponding to the areas of the field and the pond.

The purpose of this problem is to find the length of each of the three segments (the circumference, the perimeter and A). To do so, Li Ye does not directly use the sum of the three segments given in the statement. Instead, he computes one of the segments: $(diameter - side)/2$. That is,

---

[5][19.1] 今有圓田一段, 內有方池水占, 之外計地三十三畝一百七十六步. 只云內外周與實徑共相和得六百二步. 問三事各多少. 苔曰: 外圓周三百六十步. 內方周二百八步. 實徑三十四步. [19.2] 法曰: 立天元一為內方面. 以減一百七十二得 █ 為外田徑也. [19.3] 倍云數得一千二百四步. 別得是六個圓徑八個方面兩個實徑. 今將一个方面兩个實徑. 合成一个圓徑併前數. 而計是七个方面七个圓徑也. 今置一千二百四步在地. 以七約之得一百七十二步為徑面共也. 便是一个方面一个圓徑, 更無實徑也.

'*[the distance that] crosses the area*', so that he can compute the area of the circle and the area of the square. Li Ye writes directly in the first line of the procedure the new data which he finally chose to use and then explains where this unexpected data is coming from in a commentary in [19.3]. No other justification is given. However, a familiar reader will immediately notice that, the resulting reduction resembles another basic problem of the same category, i.e. Problem 11a, which uses exactly the same kind of data. That is, the data of Problem 19 are transformed in order to reduce the problem to a problem similar in type to Problem 11a.

The same type of reduction happens in the Problem 64, in which data are transformed in order to be simplified in a more general way.

Problem 64.

[64.1] Let us suppose there is one square piece of field, in the centre of which there is a ring-shaped pond. Outside the [area] occupied by water, one counts forty-seven mu two hundred seventeen *bu* of land. It is said only that the inner circumference of the water ring does not attain the outer circumference of seventy-two *bu*, and [the distances] from the four angles of the field reaching the water are fifty *bu* and a half each.

One asks how much is the inside and the outer circumferences and the side of the square field each.

The answer: The outer circumference is one hundred eighty *bu*. The inner circumference is one hundred eight *bu*. The side of the square field is one hundred ten *bu*.

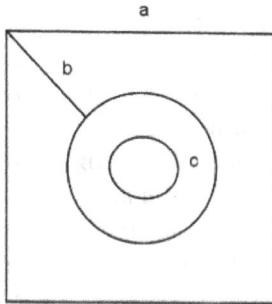

Fig. 7.5   a: square field; b: fifty *bu* and a half; c: ring pond.

[64.2] The method: Set up one *Celestial Source* as the diameter of the inner pond. First: dividing by six the difference between the outer and the inside diameters (seventy-two *bu*), yields twelve *bu* as the diameter of the water [ring]. Doubling this yields twenty-four *bu*. Adding the Celestial Source, the diameter of the inner pond, yields $\frac{2\ 4\ tai}{1}$ as the outer diameter.[6]

Here follows a transcription of [64.1]: Let d be the distance of 50.5 *bu* from the corner of the field to the edge of the pond; let A be the area of the square field (S) minus the area of the circular ring, i.e. 47 *mu* 217 *fen*. Let b be the circumference of the outer circle (B), c be the circumference of the inner circle (C), and $b - c = 74$ *bu*; let x be the diameter of C. $\frac{b-c}{6} = 72/6 = 12$. Let this quantity be named a. Diameter of $B = x + 2a = 24 + x$.

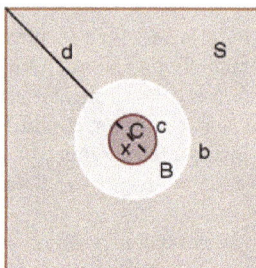

Fig. 7.6

This problem addresses two concentric circles circumscribed in a square. The statement gives the difference between the two circumferences. Li Ye starts the procedure by dividing this value by 6 to obtain the difference between the two diameters. He does not justify this transformation. Nevertheless, with this new data, he is able to compute the diameter of each circle and work with diameters only. Using the diameter transforms each circle into corresponding square areas and he then continue the computation with square areas. This procedure of transforming the area of one circle into the corresponding area of three squares is very common in the *Development of Pieces [of Areas]*. Li Ye starts to use it in Problem 11a and uses it for every circular outer field. The simplification of the data in Problem

---

[6][64.1] 今有方田一段, 中心有環池. 水占之外計地四十七畝二百一十七步. 只云其銳案: 元本作' 共' 誤. 環水內周不及外周七十二步. 又從田四角至水各五十步半. 問內外周及田方面各多少. 苔曰: 外周一百八十步. 內周一百八十步. 田方一百一十五步. [64.2] 法曰: 立天元一為池內徑. 先以六除內外周差, 七十二步, 得一十二步為水徑. 倍之得二十四步. 加入天元池內徑得 ⊞ 為池外徑.

64 reduces the problem to other problems using the same procedure. The reduction does not refer to any particular problem but to a general category of problem. The data is transformed in order to apply a procedure which is familiar to the practitioner. Because necessary procedures are provided in previous problems, there is no need for justification, but the practitioner is expected to know which problems contain the proper procedures. This tacit assumption implies that the practitioner remembers the procedures of each of the previous problem.

In both Problems 19 and 64, the transformation has the same purpose: to simplify the new problem through an association to an already known category of problem. This way of associating problems to one another is an important feature of the *Development of Pieces [of Areas]*. The logic of reducing one problem to a previously solved example and the resemblance among problems is the key to the reading. Problem 3, quoted above, presents an illuminating example, wherein a square of which one side is the diagonal of the square field given in the statement is constructed. That is, all of the square's dimensions are expanded by $\sqrt{2}$. Thanks to this expansion, the procedure can be reduced to Problem 1. In the 'meaning', Li Ye explains in detail the procedure of the expansion of dimension but does not indicate how to finish the procedure once the areas have been expanded. The reader is expected to be acquainted with the procedures already seen in Problem 1 and apply them to Problem 3.

Fig. 7.7 Problem 3

When the problems are independently solved using the Section of Pieces [of Areas] procedure, Li Ye appears to construct many of the solutions on the basis of their resemblance to previous solutions. This is particularly evident concerning problems for which the procedure is transformed to something reminiscent of Problem 1. In Problem 39, Li Ye writes the following: '*The section of areas in this problem is the same as that in Problem 1*'.[7] The two problems do not immediately follow each other. Problem 1 concerns a square field with a circular pond, and problem 39 concerns a rectangular field with a circular pond. The two diagrams indicate that the procedure is structured the same way:

Fig. 7.8   Problem 1

Fig. 7.9   Problem 39

The same remark for Problems 3 and 39 can be made for Problems 6c, 40, 42, 47 and 49. For each of these, the data is transformed in order to

---

[7] 此問與第一問條段頗同

reduce the procedure to the basic procedure of Problem 1. These problems concern various data and various combinations of geometrical shapes (circles, squares, etc.) but even though their solutions resemble each other, they do not follow each other. Why are they ordered in such a way? What rule structures the composition?

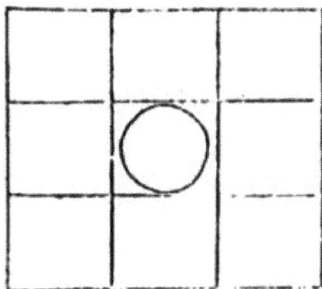

Fig. 7.10   Problem 6c

Fig. 7.11   Problem 40

At first glance, all the problems of the *Development of Pieces [of Areas]* look the same and there is no indication any specific order. Problems seem to follow each other without any particular logic. Nobody knows why Li Ye selected 64 problems and arranged them in such way. However, a brief look at the statements of the problems has already revealed that:

Fig. 7.12   Problem 42

Fig. 7.13   Problem 47

a). The problems are grouped according to the shape of the field and the pond and sub-grouped according to the construction of the distance given in the statement.

b). Some elements from the statement or from the procedure are transformed to reduce a problem using a previous one.

Thus, there is at least a basic structure of the order, and the author has seemingly classified the problems according to the given data. This order is made according to the geometric elements of the Section of Pieces [of Areas] procedure. However, a reader solving the problems one-by-one perceives the incremental development of another order. An earlier clue about this order emerged in the recommendation made by Li Ye concerning Problem 18.

Fig. 7.14   Problem 49

The study of the geometric procedure shows an analogous order implied in the solution indicated through diagrams.

Chapter 8

Analogy and Iteration: The Example of Problem 36

The process of reducing data to derive a solution via a previously solved problem expresses an analogical order. Besides the order of data based on areas and distances given in the statement, problems are also ordered by analogy. Several elements of the geometric procedure presented inside the diagram-equation direct this order. The order of problem is tacitly implied by the diagram, as seen here in Problems 36 and 45. The way of grouping problems implies a specific use of memorisation. Reductions, as seen in previous chapter, ease memorisation and reduce the amount of mental labour. The practitioner is saved the labour of repeating similar treatment in situations that are established as related through the visual quotations in diagrams. However, the role of memory is not only bound to retain procedures and geometrical patterns. Problem 36 below highlights another role for memorisation and another system of organization of problems.

8.1 Problem 36

[36.1] Let us suppose there is one circular piece of field in the middle of which there is a rectangular pond. Outside the [area] occupied by water, one counts six thousand *bu* of land. It is said only that the diagonals from the four angles of the inner pond reaching the edge of the field are seventeen and a half *bu* each. Mutually summed up together, the length and the width of the inner pond yields eighty-five *bu*.

One asks how much these three things are.

The answer: the diameter of the outer field is one hundred *bu*. The length of the pond is sixty *bu*. The width is twenty-five *bu*.

[...]

[36.10] One looks for this according to the Section of Pieces [of Areas]. Four times the *bu* of the area is added to two pieces of the square

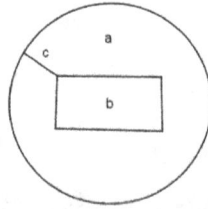

Fig. 8.1   a: circular field; b: rectangular pond; c: seventeen *bu* and a half.

of the *bu* of the sum. One subtracts twelve pieces of the square of the reaching *bu* to make the dividend. Twelve times the reaching *bu* makes the adjunct. Five *bu* is the constant divisor.

[36.11] The meaning: the two squares of the sum which are added [to four real areas] equal the quantity of eight areas [of the pond] and two squares of the difference [between the length and the width]. Inside the original [four areas], there are four empty ponds; outside, there are four areas [of the pond] and two squares of the difference [between the length and the width].

Fig. 8.2   j1–12: subtract; c1–12: adjunct; a1–2: add; B: the original [area] has.

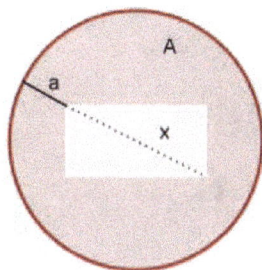

Fig. 8.3

[36.12] [To make] this dividend, one only has to complement with two squares of the diagonal of the pond. Inside four circular areas, one removes [the part] that is filled [in between] the *bu* of the adjunct, outside of the original [area], there are three squares [remaining]. Now, one adds a further two squares of the diagonal of the pond together, yielding five *bu*. Therefore, five makes the constant divisor.

8.2 Description of Problem 36

Let a be the distance from the circle to each angle of the rectangular pond, 17.5 *bu*; let b be the length added to the width, 85 *bu* and let d be their difference. Let A be the area of the circular field (C) subtract the area of the rectangular pond (R), 6,000 *bu* and x be the diagonal of the pond [Fig. 8.3].

The area $4A$ is represented by three squares for which the sides are equal to the diameter, and from which $12a^2$ is collectively removed [Fig. 8.4]. From the area $4A$, four rectangular ponds have to be removed. However, the sides of the three squares are also the diagonal of the pond added to $2a$, whereas the square that is to be removed has a side equal to b, the length and width of the rectangular pond added together. Thus, the four ponds cannot be represented inside $4A$ and cannot be removed. Li Ye cannot adjust the diagonal as he did before for the same category of problem by 'reducing' or 'augmenting by four [tenths]' (i.e. multiplying or dividing by $\sqrt{2}$ like in Problem 3). Li Ye used to transform a side into a diagonal by multiplying by $\sqrt{2}$. This procedure does not work in the present case.

Therefore, another method is used for this problem dealing with rectan-

Fig. 8.4

Fig. 8.5

gles: expressing the square of the diagonal according to the square of b. We know that two squares of the sum of the width and the length equals eight rectangles and that two squares of the difference between width and length: $2b^2 = 8R + 2d^2$ [Fig. 8.8], '*the two squares of the sum that are added [to four real areas] equal the quantity of eight areas [of the pond] and two squares of the difference [between the length and the width]*' [36.11]. Two squares of the diagonal are equal to four rectangles and two squares of the difference of the width and the length: $2x^2 = 4R + 2d^2$ [Fig. 8.9]. Li Ye expresses this as the following: '*Inside of the original [four areas], there are four empty ponds; outside, there are four areas [of the pond] and two squares of the difference*' [36.11]. With these two sentences, Li Ye expresses b in terms of x, that is, if $2b^2 = 8R + 2d^2$ and $2x^2 = 4R + 2d^2$, then $4R = 2b^2 - 2x^2$.

The initial subtraction of four rectangular ponds from the three squares representing four circular areas is represented by the four empty rectangles in [Fig. 8.9]. These can thus be replaced by adding two squares in which the side is the diagonal of the rectangle, x, instead. Li Ye writes '*one only has to complement with two squares of the diagonal of the pond*' [36.12]. The problem then proceeds normally: to read the area in terms of the unknown, one has to identify $12ax$. Inside are three squares whose side is the unknown [Fig. 8.5]; the two other squares representing $2x^2$ are stacked on the diagram [Fig. 8.6], thus giving $5x^2$. We thus read $4A = 12a^2 + 12ax + 3x^2 - 2b^2 + 2x^2$, which can be visualised as $4A - 12a^2 + 2b^2 = 12ax + 5x^2$.

Table 8.1

$4A - 12a^2 + 2b^2$	dividend
$12a$	adjunct
5	constant divisor

In Problem 36, the 'meaning' describes a figure which is not represented in diagram. This figure is made of two gnomons. The elements required to solve Problem 36 are not visible. Nonetheless, the reader has already dealt with gnomons earlier in the *Development of Pieces [of Areas]*. The figure necessary to understand the origin of the terms of the equation for Problem 36 is given in Problem 32:

However, the justification of the procedure is given only in Problem 33, where the 'meaning' states: '*The sum of four ponds with the square of the difference [between the length and the width] which is subtracted there is exactly one sum of [the length and the width] by itself*'.[1] Thus, with Problems 32 and 33 in mind, it is not difficult to understand Problem 36.

Fig. 8.6

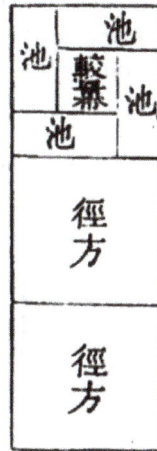

Fig. 8.7 Problem32

There is no coincide if precisely this operation is absent from Problem 45. In the Section of Pieces [of Areas] procedure of Problem 45, Li Ye writes

---

[1] 四池并所减底个較冪, 恰是一个和自之.

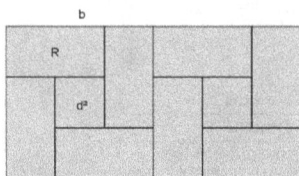
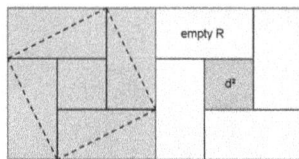

Fig. 8.8   Fig. 8.9

only [45.2]:

[45.2] *One looks for this according to the [procedure of the] Section of Pieces [of Areas]. One relies on the previous pattern, which implies that it is not necessary to draw [another diagram]. One only sets up the real area and breaks it into four pieces of small rectangular fields. The side of the pond makes the difference [between the length and the width of the rectangle]. The side of the outer square field makes the sum [of the width and the length of the rectangle]. The reaching bu on the diagonals makes the hypotenuse. The problem is then precisely [as follows]: as the pond stands right in the middle of the square field, according to the method, one can look for [the unknown].*[2]

The following diagram is provided in the statement of this problem: [Fig. 8.10]

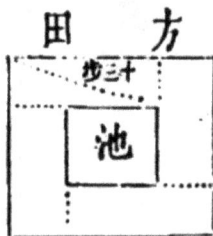

Fig. 8.10   Problem 45

This problem does not have the usual sentence describing the coefficients of the equation, and there is no specific diagram for the Section of Pieces [of Areas]. None of the Qing dynasty commentators count this as a loss, nor

---

[2]依條段求之. 只據前式, 便是更不須重畫也. 只是將見積„ 打作四段小直田. 以池面為較. 以外田方面為和. 以斜至步為弦. 然此問惟是其池正在方田中心, 可依此法求之.

have they added either corrections or supplements to this problem. Indeed, the diagram given in the statement is sufficient to identify the equation because it is the same as the diagram for the Section of Pieces [of Areas] procedure would be. The coefficients of the equation can be found in turn on the basis of previous problems.

Despite the lack of description of the coefficient, reconstructing the equation in modern terms is possible:

- Let s be the side of the outer square.
- Let w and l be the width and the length of the rectangles (R), respectively.
- Let x be the side of the interior square, $x = l - w$.
- $s^2 = 4R + x^2$
- $a^2 = 2(w \times l) + (l - w)^2$
- $a^2 = 2R + x^2$
- $2a^2 = 4R + 2x^2$
- $4R = 2a^2 - 2x^2$
- Thus $s^2 = 2a^2 - 2x^2 + x^2 = 2a^2 - x^2$
- $A = s^2 - x^2$
- $A = 2a^2 - x^2 - x^2$
- $A = 2a^2 - 2x^2$
- The equation is: $2x^2 = 2a^2 - A$.

The equation in Problem 45 is encapsulated in a single diagram [Fig 8.11], which confirms that diagrams can be a sufficient indication to the reader of the procedures used to set up the equation.

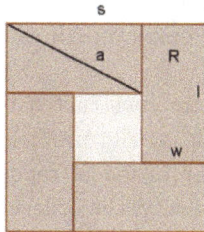

Fig. 8.11

Solving a problem is thus merely finding a path through the previous procedures. There is no narrative description of the sequence of operations. The diagram 'tells' the practitioner what they are supposed to do. To solve the sequence of Problems 32, 33, 36 and 45, the practitioner must extract and gather similar examples from among several problems to solve the next one. The order of operations is tacit and is expected to be learn by the reader is in the course of solving problems.

The structure of the *Development of Pieces [of Areas]* relies on memorisation of all of the previous problems. To find the solution for Problem 64 (or for Problems 33 to 45 here), a path must be drawn through the sixty-three previous problems to find which of the procedures are required for the geometric solution. It appears that the diagrams were not only visualised but also memorised. However, memorisation is not merely a copy and paste function, but is it a way of verifying of knowledge or a way of preserving information which could not be written. Here, memorisation is a strategy for a deeper understanding through the impression left by visualisation. Memorisation creates a general and abstract framework for understanding. It is a practice undertaken deliberately, not simply taken for granted. The technique is systematic in the sense that it produces clearly specified thoughts. It does not rule out spontaneous or even random, creative elements. It is continuous; meaning that the intentional activity undertaken is either continued or repetitive. This description is close to the description of the technique of meditation by Eifring [Eifring (2016), 5]. According to him, meditation is a technique in the sense of being a deliberately undertaken and systematic practice involving continuous (i.e. repetitive or continued) activity aimed at producing certain effects at least partly by means of universal mechanisms [Eifring (2016), 17]. He showed that visualisations are often explicitly content-oriented and have for their purpose the direct realisation of truth. They are employed to change mental contents such as thoughts and images.

The geometrical figures are constructed to recall other figures, rather than to show what is unique to the problem in question. This may be the reason for the strong impression of repetition in the *Development of Pieces [of Areas]*. On the contrary, the ordering by resemblance shows that the figures are actually meant to underline what is common between questions. Li Ye's 'meaning' describes only a few of the specifics required for the solution because his objective is to attach the new problem to a family of problems which are already understood. Diagrams are an exploratory model. The practitioner acts like a cartographer. He describes borders, and demarcates the passages from one figure to another. He places landmarks, delimits, constructs series, and manipulates representations. In the end, the explorations of the practitioner constitute a dynamic process until a unity emerges from these representations. He thinks of transformations and exchanges with simple geometrical figures, such as squares or circles. With this operational apparatus which is easy to represent and can work at several levels, the practitioner animates the figures. That is, he or she concretely introduces becoming into being, generality into diversity. The exhibition of the general makes sense only if it represents a multiplicity of cases. The expression of generality is presented in the form of an enumeration of solutions displayed analogically, where generality is explored through visualisation and memorisation of validity of procedures. Here, harmony seems more valued than conciseness or simplicity.[3]

All the problems in the *Development of Pieces [of Areas]* are ordered according to their resemblance to other problems and each solution is a combination of procedures given for solutions of previous problems. The problems are not ordered according to their complexity, their degree of difficulty, or their use of new procedures but according to a web. This web is arranged according to a systematic correlation of the procedures involved in the Section of Pieces [of Areas] procedures. Each problem shows a possible combination of the transformations involved in the procedure. The structure is made through analogies, which are made evident through the construction of the diagrams. The structure of the text that is thus revealed conveys its mathematical meanings.

---

[3]See Knoblock [Knobloch (2016)] on Leibniz on this point.

Chapter 9

The Web of Dependencies: The Example of Problem 14

Problems in the *Development of Pieces [of Areas]* are ordered according to the possible combinations of the operations presented in Chapter 4. These combinations are applied to various figures and data. It seems that the Section of Pieces [of Areas] procedure explores the entanglements of several sets of combinations:

(1) A finite set of combinations of types of figures (circles, squares···). (Chapter 7).
(2) A finite set of combinations of types of data (Area and distance given in statement). (Chapter 7).
(3) A finite set of combinations of transformations on geometrical areas. (Chapter 4).
(4) A finite set of mathematical objects (fractions, half-areas, powers, full or empty areas···)

This construction is organised around an architecture motivated by analogy and reduction. For instance, the first chapter of *Development of Pieces [of Areas]* is organised like a mirror, as described in [Table 9.1]. The transformations in Chapter 1 are summarised as types A–F:

A: to remove the corners;
B: to pile and unstack the adjunct;
C: to compensate areas;
D: to expand areas;
E: to multiply areas by parts;
F: to cancel areas

The first chapter of the *Development of Pieces [of Areas]* contains all the basic operations. This chapter supports to introduce these operations,

Table 9.1   Order of Transformations in Chapter 1

Type of transformations.	Problems of the shape 'square field with an inner circle.'	Problems of the shape 'circular field with an inner square.'	Problems of a different shape.
A	Problem 1		
B		Problem 11a	Problem 21 operation F
B + D		Problem 11b	Problem 22 operation F
B + C	Problem 2	Problem 12	
A + D	Problem 3	Problem 13	
B + C + D	Problem 4	Problem 14	
B + E	Problem 5	Problem 15	
A + E	Problem 6	Problem 16	
A + C + E	Problem 7	Problem 17	
B + E, 10^5	Problem 8	Problem 18	
B + E, 10^6	Problem 9	Problem 19	
B + E, 10^7	Problem 10	Problem 20	

how they can be combined and how they can be applied to various figures and data types, whereas the other two chapters present sophisticated forms of these operations. The first chapter does not contain the transformation AA or BB. In this first chapter, for the figure of a circle inside a square (Problem 1 to Problem 10) and its opposite, a square inside a circle (Problems 11 to 20), the problems are coupled according to similar sequences of transformations: 6 with 16, 7 with 17 ···10 with 20. See [Table 9.1].

The first problem involving a circular field with an inner square pond, Problem 11, is composed of two problems with different statements. The two problems can be separated into 11a and 11b. Problem 11a does not use the basic procedure but deals directly with the stacked adjunct areas, as Problem 2. Problem 11b, on the other hand, addresses the procedure of expanded areas presented in Problem 3. Problem 11 thus contains procedures A, B and D, exactly as if the procedures of problems 1, 2 and 3 were grouped together. Coincidentally, 11a is identical to the one presented by Yang Hui in the *Fast Methods of Multiplication and Division*. Then, as with Problems 2 and 12, all the other problems correspond to each other according to the pattern noted above. Problems 8 to 10 are treated in exactly the same way for transformations B and E, the only difference being the given quantities, which increase by one power each time: the dividend is of the order of 10^5 in Problem 8, the order 10^6 in Problem 9, and the order of 10^7 in Problem 10. The same phenomenon occurs in Problems 18,

19 and 20. Problems 21 and 22 concern other figures (three squares for Problem 21 and a triangle inside a square for Problem 22) and introduce a new operation which mirror the two problems collected in Problem 11, etc. The sequence of procedures is organised in the same way.

    The organisational architecture of the *Development of Pieces [of Areas]* can be represented and clarified by means of graphs. Problems in which the solutions use the same transformations are linked together. Also, problems which belong to the same family of shape or involving same objects (powers, fractions, etc.) have been grouped together when possible. Family of objects are marked in green in the graph. Several graphs related various series (or sub-series) of problems appear. These graphs are not totally independent. The first two operations of the *Development of Pieces [of Areas]* illustrated by the first two problems function like opposites. Operation A occurs in directly removing areas, whereas operation B occurs in displaying areas. These operations constitute two opposite categories of problems to which other problems are related. Either a problem will correspond to type A or type B. Some problems are working like doors. Thus, a given problem belongs to one architecture but shares striking similarities with problems from other architecture. Architecture are visually represented by graphs in Appendix A of this book. For instance, Problem 5 of graph 1 may be related to Problem 26 of graph 2. These points of contact are marked by the mention "link to web" in the graph.

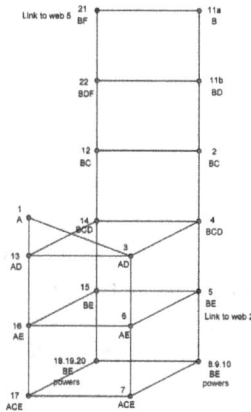

Fig. 9.1   Graph 1 of Problems 1–22

Table 9.2   Order of Transformations Chapter 2: Problems 23 to 30

Problem	Figure	Transformations	Equation
23	Square + circle	A + E	$14A = 14a^2 + 28ax + 25x^2$
24	Square + circle	BB1	$a^2 - A = 2ax - 1.75x^2$
25	Square + circle	A + E	$176A = 11a^2 + 22ax + 25x^2$
26	Square + circle	BB2	$3a^2 - 48A = 6ax - 7x^2$
27	Square + circle	A + E	$14A = 11a^2 + 22ax + 25x^2$
28	Square + circle	A + E	$176A = 14a^2 + 28ax - 25x^2$
29	Square + circle	BB3	$4a^2 - 48A = 8ax - 7x^2$
30	Square + square	BB4	$21a^2 - 28A = 42ax - 43x^2$

This structure is quite obvious for the first twenty problems, but things get complicated for the remaining forty-four problems. For example, the series of Problems 23 to 30, which concerning a square field next to a circular field, or two circles next to each other, cannot be related to the mirrored graph of Chapter 1. This sequence of problems presents another combination of iterations and alternations which contains only the three basic operations, i.e. A, B and C. Nonetheless, operation B is sophisticated and interact with various objects, as if the author wanted to exhibit the possibilities of this operation.

[Table 9.2] shows there are two iterations: alternate iteration of the operations A + E and B and iteration of the polynomials resulting from the operations A + E. If $14A$ is renamed 'α'; $14a^2 + 28ax + 25x^2$ as 'β'; $176A$ as 'γ' and $11a^2 + 22ax + 25x^2$ as 'δ', then the equations of problems 23, 25, 27 and 28 display another alternation in the components of polynomials:

- Problem 23: α β;
- Problem 25: γ δ;
- Problem 27: α δ;
- Problem 28: γ β.

The remainder of Chapter 2 (from Problems 31 to 42) combines of two groups of transformations. Problems 31 to 36 contain gnomons in their solution and they operate with B and A. Here, the operation A is complicated by its application to increasingly sophisticated objects. Only Problem 31 contains operation F which is systematically involved in the rest of the chapter (Problems 37 to 41). Problems 39 and 42 can be used as a link to the graph of the next chapter. There are two graphs in Chapter 3: one for the Problems 45 to 54 and the other for the Problems 55 to 64. However, it is not possible to display the graphs satisfactorily in two-dimensions because of their complexity. Here, the author contends with

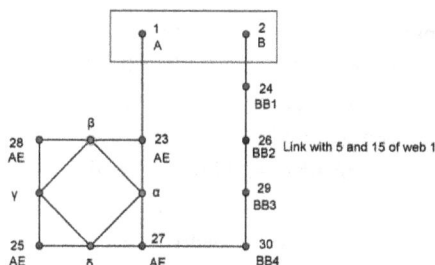

Fig. 9.2   Graph 2 of Problems 23–30

the sequences of operation, the arithmetic and geometry of the solution, the figures and data from the statement of the problem and analogies to other problems. These influences reveals why the series in [Table 7.1] and [Table A.2] could not manifest a visible order in Chapter 3. At the end of the treatise, the last six problems are solved with very simple operations and display very simple diagrams, but they have more complex objects resulting from the addition of polynomials. The only way to solve these problems is to know how they refer to other, previous problems. These three last graphs [Fig. S.2], [Fig. S.3], [Fig. S.4] are provided in Appendix A.4. Following the links in the second graph, these five graphs can be incorporated into the general graph [Fig. S.4] in Appendix A.4.

However, this graphic representation is not satisfactory. Graphs or tables naturally pertain to objects which exhibit some similarity in the diagrams, some relationship among problems. Unfortunately, these graphs are nothing but didactic tools. Relying on these graphs to understand the structure of the *Development of Pieces [of Areas]* means neglecting an essential dimension: movement. Graphs are static while the architecture, the web of problem, demands a never-ending review. The practitioner visualises transformations to create a diagrammatic equation for one problem but also connects these transformations with other problems. Movements within other problems, the relationships between mathematical objects, filled or empty area, all occupy the mind as it travels in a maze for which walls are constantly moving. The Ariane's thread to this maze is the geometrical construction of algebraic equations. Engaging in solving one problem is to disengage another problem.

This perpetual disconnection leads from one mathematical space to another, to a new plan, a level of understanding, reference to another problem, from expansions to contractions. This mental gymnastic introduce multiple dimensions, and imposes changes of perspective. The practitioner cannot settle on a plan, or stick to one referent. The mind must continually 'jump' from a level to another. Therefore, the practitioner must also pass from one dimension to another in order to make structural identifications, like transparent superimpositions. This mental travel creates knowledge through relation of equivalences between two or more terms, polynomials, and diagrammatic equations. Identities recall each other through terms of discourse shaped as phrases and visual quotations. This system of ellipsis and infinite analogy establishes a unity for the procedure. The global visualisation of the web of problems forms an impression of the generality of the procedure. Impression is a strong effect on the intellect, an immediate effect of an experience. The recognition of the generality results from mental practice of exercises of visualisation. The practitioner intuits why the procedure is true. Unity is seen — literally seen — behind diversity. Here the question becomes what creates the feeling that reasoning is valid. Logic and cognition are related. The mathematical unity manifests through a wide range of examples, which seemingly destroy unity. However, it can be realised precisely because diversity negates unity. There is no unity without diversity, and this general vision appears once the practitioner reaches Problem 64, having solved each problem individually.

It is not possible to represent the web of relationships among the 64 problems and their transformations. The danger lies in its reification. The web is something alive, multifaceted, and moving. Human spirit, or imagination, must create terms, points of intersections, complex relations. Imagination must visualise the web of association, organise its pieces, reassemble them into series, gather results, dissociate old connection, undo patterns and reform all relationships. The mind must interpret and translate. Once the practitioner sees why procedures to solve problems are true, he understands what validates a move from a particular proof to a general conclusion.

The order of the problems is not evident at first glance but appears only through solving the problems individually. The combination of procedures is so intricate that modifying this order is difficult; for example, Problems 12, 13 and 14 follow an evolution of the procedure presented in Problems 2, 3 and 4. These problems cannot be re-ordered and the procedure cannot be modified without destroying the meaning. In Problem 14, for instance, the procedure relies on three operations: unstacking the adjunct, compensating areas and expanding areas, like Problem 4. If Li Ye had changed the procedure as he recommended, then two of the operations would be missing and the analogy with Problem 4 would be lost. Thus, Li Ye chose to add a simple recommendation rather than modify the procedure used in Problem 14. He chose to preserve the analogy. Later, he does the same thing in Problem 18, and he recommended other changes in procedure in Problems 44, 52 and 56. The order of problems is based on the order of procedures, and Li Ye does not want to alter either of the orders, lest he destroy the correspondence. He points out that the procedures could be improved, but refuses to correct or re-order the problems.

Based on this observation and on the fact that the vocabulary for a negative coefficients and diagrams is different between the procedure and the commentary, Li Ye may be hypothesised as not having devised this organisational structure himself but attempted to preserve a pre-existing architecture. This conclusion implies that Li Ye borrowed more than just the twenty-three Old Procedure problems from the *Collection Augmenting the Ancient [knowledge]*, from which he took his inspiration. The 64 statements and their geometrical solutions are probably as old as the *Collection Augmenting the Ancient [knowledge]*; consequently, the role of the Old Procedure in the *Development of Pieces [of Areas]* merits reconsideration. Perhaps Li Ye intended such in the preface when he wrote: '*[For instance], a book entitled Collection Improving the Ancient [Knowledge] was compiled recently with reshaped) [solutions to geometric problems of] rectangles and circles. It is indeed an equivalent of Liu Hui and Li Chun-Feng. However, I detest its reserved style, and hence added detailed diagrams of how to reshape the Sections of Areas*'. Li Ye added diagrams to 64 of the problems without changing their underlying order. His purpose was not to make things easier and he was not simplifying the procedures which he presented in the *Sea Mirror of Circle Measurement*. He did not even mention the procedure of the Celestial Source in his preface. Instead, he transmitted and clarified another procedure. The *Development of Pieces [of Areas]* is not a textbook on the procedure of Celestial Source with a collection of simple problems; it is the preservation of an ancient organisational structure.

How ancient is this organisation? The *Collection Augmenting the Ancient Knowledge* has been assumed to be composed in the eleventh century. However, its title suggests that its content is older. It is impossible to certify the origin of the structure and guess the history of its composition. Moreover, it is possible to relate this practice of alternation, iteration, transformation, and visualisation to a more general dimension of the intel-

174 The Empty and the Full: Li Ye and The Way of Mathematics

lectual history of China [Knuth (2013)]. Based on the available elements, the inclusion of Song Dynasty mathematics in Song Dynasty intellectual milieus would be highly speculative. Nevertheless, it is possible to open a limited speculation. There is place for a future chapter on history of mathematics here. One of the oldest lists of binary n-tuples can be traced to the *Book of Changes* (*Yi Jing* 易經). The book consists of sixty-four chapters, wherein each chapter is symbolised by a hexagram of six lines, each of which is either - - (*yin* 陰) or — (*yang* 陽). Traditionally, it is said that Zhou Wen Wang 周文王 or King Wen arranged the hexagrams in their sequence while imprisoned by Shang Zhou Wang 商紂王 in the twelfth century BC. However, it has been notoriously difficult to assign a date to the sequence and there is no primary evidence that anyone compiled such a list before third century BC. There are several orders of the hexagrams. A different arrangement originated in the Song Dynasty. The binary sequence was named in honour of the mythic culture hero Fu Xi 伏羲. It is believed to be the work of the scholar Shao Yong 邵雍 (1011–1077 AD). Of the two hexagram arrangements, the King Wen sequence is, however, of much greater antiquity than the Fu Xi sequence.

Similar to the problems of the *Development of Pieces [of Areas]*, there are 64 sections. In the King Wen arrangement, hexagrams occur in pairs. Each diagram is immediately followed by its top-to-bottom reflection. For instance ▰▰, the first is pure *yang*, the second is pure yin. In the *Development of Pieces [of Areas]*, data alternate in the statement of the problem and the operations A and B are used alternatively throughout the sequence of problems.

The enumeration of objects and operations, their transformations, the appearance of patterns within the order, and the arrangement and combination of objects, draw the *Development of Pieces [of Areas]* closer to combinatorics and magic squares than to algebraic solutions of quadratic equations. After all, as Li Ye stated in his preface:

'*On the other hand, contemporary mathematicians, who do not necessarily study as comprehensively as Liu Hui or Li Chunfeng, are narrow-minded and short-sighted. Instead of making it clear, they prefer rendering it as implicitly and intricately as possible in order to make mathematics appear opaque and obscure. They prevent even a glimpse of its simulation being caught by others. Otherwise, some of them opt to deal with merely the basic and well-known part that is not worth looking into. Consequently, the methods of the ancients Xuan Yuan and Li Shou along with combination*

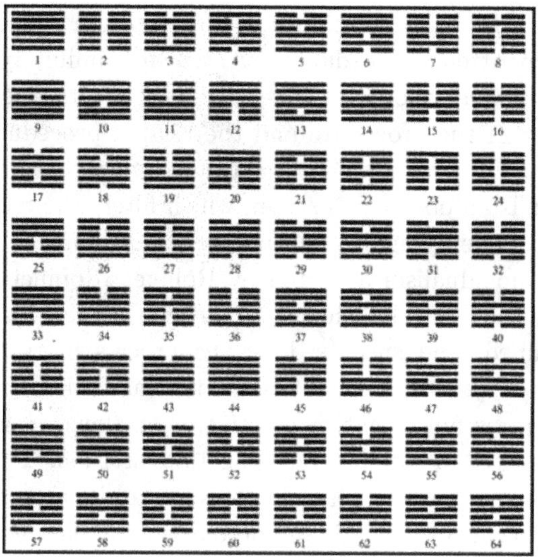

Fig. 9.3 King Wen Sequence

and alternation of numbers by three and five (三五錯綜) become something with which everyone in the town can be self-satisfied. It is such a pity that they actually know just as much as ignorant villagers.'[1]

The terms '*combination and alternation*' (*cuozong* 錯綜) ([Bréard (2012), 3]) are borrowed explicitly from the *Book of Changes* which is translated as '*By three, by five, through the transformations; altering and combining the numbers.*'[2]

This discourse by number shares fundamental features with Taoism. Dao is ineffable and the mystical experience unspeakable. When Taoists are confronted with the universal problem of transmission and translation into words of something escaping language, they paradoxically rely on language to transmit and educate. As an indispensable obstacle, language must be recreated. Diagrams are silent signs which contain the surplus of meaning. They can capture unspeakable meanings. It is not enough to recall the value of silence. But silence must be introduced to the discourse. The essential intertextuality of silence produces meaning, as discourses and diagrams

[1] 今之為算者，未必有劉、李之工，而褊心跼見，不肎曉然示人，惟務隱互錯糅，故為溟涬黯黮，惟恐學者得窺其彷彿也。不然，則又以淺近犄俗，無足觀者，致使軒轅隸首之術，三五錯綜之妙，盡墮於市井沾沾之見，及夫荒郙下里，蚩蚩之民，殊可憫悼。

[2] 參伍以變，錯綜其數。

work together. The textual practice is reinvented in silence. There is a fundamental unwritten yet readable text, which is infinitely interpretable. That text is the universe, or Dao. The successful practitioner perceives the infinite plurality of the procedure and the related procedure as a picture of the universe. For someone who wants to signify the movement of the world, language is an obstacle, a screen which filters the truth. Therefore, the discourse is made by figures and binary oppositions.

Binarism is not dualism as noted by Robinet [Robinet (1979)]. The polarity should not be abolished. On the contrary it must be integrated, maintained, and surpassed by the process of transformation as performed by the practitioner. This fundamental binarism recovers every polarity: empty/full, negative/positive, short/long, inside/outside, diminished/augmented, up/down, multiplication/division. The purpose is to play with this binary, to make it work, to show all its aspects. The practitioner, after following every path and detour is truly dazzled. Equivalences are superimposed, contradict each other, cross, converge, and diverge by alternation, complement, contrast, and change of element. Here, language is not descriptive, but operational.

Li Ye identified operations that correspond to the fundamental of the *Book of Changes*. The composition of the *Book of Changes* manifests how changes develop as the interaction of two opposed, but complementary, principles: The Yin and Yang. Similarly, procedures, which embody changes within mathematics, indicate the crucial part played by inverse operations. They are crucial building blocks for producing procedures. Operations and objects are inverse of one another. Thanks to them, the practitioner can work on the generation of procedures, hence Changes.

Diagrams and their organisation are structures. A diagram makes sense only by comparison with other diagrams. They are immanently tied to their background — to a nexus of meaningful relations among objects within an organisation structure. Because mathematical objects are inextricably embedded within meaningful relationships, each object reflects another. The practitioner tacitly experiences all the perspectives upon that object coming from all its environment, as well as the potential perspectives which that object has upon the beings around it. The perception of the practitioner is ambiguous, founded upon the understanding of the structure and its meaning. Perception has an active dimension. Each diagram is a mirror of all other. The structure is not a collection of units, but of oppositions. This structure is a reality wherein the whole defines the parts, wherein parts have reality by differentiation only. In this structure matters first (and may be only) relations.

Only the act of visualisation can create meaning. Only differences and hence their relations create significances. A web of diagrams without words is thinking in construction. Because the process is always under progress, it can only be caught by movement.

Conclusion to Part III

An investigation of the structure leads to a deeper understanding of the problems and procedures and to a different mathematical meaning, which remains to be unveiled. Chemla's experience with the *Nine Chapters* led her to believe that that classic is also subject to the same phenomenon of reference between problems [Chemla (1997a)]. Some studies have already underlined the importance of analogical reasoning in China, such as [Volkov (1992)], which demonstrated its role in the commentary by Liu Hui to the *Nine Chapters*. Liu Hui prescribes 'mending the void with the excess' (*yi ying bu xu* 以盈補虚) as a procedure to solve problems in the first (Problems 25–26) of the *Nine Chapters*.[3] This process is used for the computation of the area of an isosceles triangle, which is part of a figure called *gui tian* 圭 田.[4] Half a triangle, ABC, is applied as a 'missing' part ADE to complete the rectangle ADEC.[5]

Liu Hui does not provide the details of the transformation and there is no reference to diagrams. Volkov suggested that the diagrams could be imaginary and that '*the reader is expected to be able to perceive objects supposed to be obtained even before the operations were actually performed*' [Volkov (2007), 441], which would explain the absence of references to any figure in this case. The earliest reference to the operation of 'mending the void with the excess' is in a diagram found in Yang Hui's *Fast Method of Multiplication and Division*. Volkov wondered '*whether Yang Hui had access to the same tradition of diagrammatic representation as did Liu Hui, or whether he had imitated an alternative (or more recent) tradition, probably*

---

[3][Volkov (2007), 68–71]; [Volkov (1994), 134]; [Volkov (2007), 440]; [Chemla and Guo (2004), 138]. The expression is translated by Karine Chemla as 'avec ce qui est en excedent, on comble ce qui est vide' (with what exceeds, one completes what is empty).

[4]For the interpretation of the shape of *guitian* see [Volkov (2007), 68, note 23].

[5]See [Chemla and Guo (2004), 137–138] for more details.

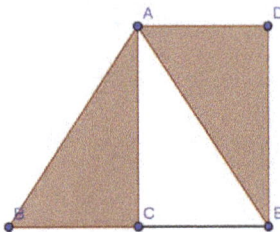

Fig. 9.4

aiming at reconstructing Liu Hui's original diagrams' [Volkov (2007), 446]. The transformation described by Li Ye recalls the strategy prescribed by Liu Hui and represented by Yang Hui. In the *Development of Pieces [of Areas]*, 'empty' or 'full' areas are pasted, removed, restored and completed. These three authors, Liu Hui, Yang Hui and Li Ye, approached procedures as transformations, wherein operations are opposite and complementary ('empty' (*xu* 虛) 'compensating' or 'mending' (*bu* 補) in *Development of Pieces [of Areas]*). It is thus no coincidence that Liu Hui is named in the preface of the *Development of Pieces [of Areas]*, as there appears to be a connection.

However, this reference to Liu Hui in Li Ye's preface could also be justified by their similar use of analogous reasoning. Liu Hui uses the strategy of 'mending the void with excess' on two different figures to transform them. He starts with a triangle and then shows how a trapezoid is transformed into a rectangle. He then uses this method as a model for three-dimensional cases. That is, his argument is based on the comparison of two situations which resemble each other in a certain aspect and consists of demonstrating the general case of a situation through examples. Volkov [Volkov (1994), 135] wrote that: '*Chinese mathematical texts have a way of presenting data which is based on the intuition of analogy of objects or methods rather than on the idea of a deductive inference. The principal rhetorical device was a transferable example, or a model, that is, an example, usually simple, which could be transferred by analogy into the given domain to generate a sound (and, usually, complex) result'*. In another example, Liu Hui introduces the general method of computing the volume of a pyramid by choosing a pyramid in which the base is square and the width, length and height are equal, which was considered as a model [Volkov (2007), 67]. Here, the general case is established by means of the transformation of a figure and by the verification of the method used. The common practice was thus to use analogies for demonstration and their generalisation.

According to Chemla [Chemla (1997b), 201–202], the operation of 'mending the void with excess' is also a schema of demonstration. In his commentary, Liu Hui justifies a computation through its link to a general transformation. The transformations are described in terms of 'excess', 'void' and 'compensation' and are used identically for different geometrical transformations. A *rapprochement* is made between demonstrations of divergent procedures through recurring underlying operations. However, Chemla does not analyse the operation in the context of analogical reasoning and justifies the absence of diagrams differently. Chemla reported different occurrences of the same rhetoric in the context of Economics and in hexagram fifteen of the classic *Book of Changes*.[6] She deduces thence that the practice of reflection on transformations was not confined to the mathematical context. Philosophical reasons probably led to the focus on transformations in mathematical procedures. According to Chemla, Liu Hui appears to conduct a general reflection on Changes, embedded in a mathematical experience.[7]

She showed that a quest for the greatest generality directed the mathematical work of Liu Hui. She observed that that the relation between the statement of problem and its solution is not obvious [Chemla (2006b)]. In the *Nine Chapters*, the problem is not solved by a random procedure, but by the most general procedure possible. Therefore, it is possible to establish a particular problem and its following procedure as a general statement. The demonstration of the correctness of algorithm is oriented toward the establishment of general operations. She identified many other indications that generality was a major epistemological value of the *Nine Chapters*, while the book itself includes no explicit discussion of generality. General operations are presented on an object with particular dimensions, but they do not use the singularity of the dimensions. There may be no abstract operation, but there is still generality. In Liu Hui's commentary, the generality is sought from operations, not from a statement of problem.

---

[6]See also [Chemla and Guo (2004), 1025].

[7]Volkov [Chemla (1997b), 46] also suggests that a large part of what the modern historian reconstructs as 'traditional Chinese science' actually was a subset of a 'complex network of various social and cognitive activities among which religious, philosophical, and mystical teachings and magical practices played an extremely important role'.

Problems in the *Development of Pieces [of Areas]*, like in the *Nine Chapters*, cannot be reduced to simple statements requiring solutions, such as 'problem solving' for school [Chemla and Guo (2004), 32]. Rather, as Chemla [Chemla (2000)] showed, problems can be used to interpret the operations of an algorithm. In the *Nine Chapters*, problems offer contexts wherein operations necessary for the demonstration of the procedure can be interpreted. As previously noted, in the *Development of Pieces [of Areas]*, as in other Chinese mathematical texts, the answers to the problems are given immediately, whereas the way to solve the equation is never given. The answer is not the goal of the problem. Thus, in the *Development of Pieces [of Areas]*, problems are used for their argument, and the category of 'algebra' takes the shape of a list of problems. But the *Development of Pieces [of Areas]* presents a systematic exploration of a procedure through combinations of data or problems.

Several common features unity Liu Hui's commentary and Li Ye's work, such as analogy, transformation, geometrical demonstration, generalisation, and context for the problems. There are, however, differences between Liu Hui's commentary and the *Development of Pieces [of Areas]*. In the commentary to the Nine Chapters, one procedure and one case is sufficient for a demonstration, and it can be used analogous for several different problems, whereas in the *Development of Pieces [of Areas]*, there are sixty-four problems concerned by the same procedure. The practice of problems evolved into an articulated system of understanding the mathematical validity. The practice of problems, analogy, and generality in the *Nine Chapters* and the *Development of Pieces [of Areas]* differ slightly. Here, it exhibits a way of classifying equations through analogy.

Chemla [Chemla (2005)] has already observed an evolution of generality from the first to the third centuries in China, that is, between the composition of the *Nine Chapters* and the *Gnomon of the Zhou* and their respective commentaries. There were surely other changes afterwards. The *Development of Pieces [of Areas]* exhibits one new phase of this evolution. Sixty-four problems give more possibilities to express of Changes. In the *Nine Chapters*, generality was expressed through the choice of procedure. In the *Development of Pieces [of Areas]*, the multiplicity of procedures and their possibility of combination express generality. Generality is not an assumption or the result of simplification and conciseness. The practitioner can obtain generality only after the effort of exploration. In the *Development of Pieces [of Areas]*, if one figure is sufficient to derive an equation, it is not sufficient to create an impression of generality. Generality expresses itself though combination of multiple figures. Likewise, the diagrams are not pre-established or given. The practitioner had to construct them. In Liu Hui's commentary, as well as in the commentary to the *Gnomon of the Zhou*, "the figure was the basis for establishing in a uniform way the correctness of several algorithms" [Chemla (2005), 145]. In the *Development of Pieces [of Areas]*, a unique web articulates all figures and algorithms. That web commands the greatest number of possible algorithms. Generality thus expresses itself through the organisation of the order of problems, which appear after mental visualisation of transformation of diagrams. The sophistication and expansion of this web communicates a broader idea of generality.

Chapter 1, Section III already noted that each sentence of the Celestial Source procedure retranslated each polynomial in term of its geometrical signification. This repetition links the algebraic and the geometrical procedures. Li Ye does not detail the computations of any of the algorithms, but he systematically determines the signification of the coefficients obtained after sequences of operations. This signification is introduced by the character 'as' (*wei* 為) in both procedures or 'meaning' in the Section of Pieces [of Areas] procedure. We already know that the objective of the algorithm is to solve neither the problem in general nor the equation in particular. It may be that the algebraic Celestial Source procedure serves to establish the proof of the correctness of the geometrical algorithm. The statement of the algorithm is in fact part of a demonstration [Chemla and Guo (2004), 335, 461, 672]. Thus, it is more important to find a meaning in the algorithm than it is to find a numerical root. It would make more sense to imagine that Li Ye wanted to preserve an ancient structure that explored the construction of equations with negative coefficients, refresh the old obscure geometrical procedure by providing new diagrams, which were missing and prove its correctness with a new procedure.

There is thus abundant opportunity for further study, as this study of the *Development of Pieces [of Areas]* initiates a study of relations between the milieus of literati in China. At least, it has now been shown that the *Development of Pieces [of Areas]* is not merely a popular mathematical book which introduces the Celestial Source procedure for beginners. In fact, the objective of the *Development of Pieces [of Areas]* is to capture the unity behind diversity. This practice which is to be related to the specific milieu shows how the expression of generality can differ from expectations. The comprehension of why things are true is bound to ways of convincing which depend on cultural expectations. The *Development of Pieces [of Areas]* is far from being a practical mathematical textbook. On the contrary, its roots are essentially philosophical. It aims to look for the elements of nature. Li Ye develops a general reflection on Changes within mathematics, which bases itself on the observation of diversity of procedures.

Problem 3, 36, 14

Problem 3

第三問

[3.1] 今有方田一段, 內有圓池. 水占之外計地一萬一千三百二十八步. 只
云從外田角斜至內池楞各五十二步.
問內徑[8]外方各多少.
荅曰: 外田方一百二十步. 內池徑六十四步.

Fig. 9.5 a: 方田; b: 圓池; c: 五十二步.

[3.2] 法曰: 立天元一為內池徑. 加倍至步得 ⊟ 為方斜.

[3.3] 以自增乘得 ⊟ 為方斜冪, 於頭. 其方斜上本合身外減四. 今不及
減便是寄一步四分為分母也. 今此方斜冪乃是變斜為方面. 以自乘之
數, 又別得是展起之數也.
[3.4] 又立天元為池徑. 自之, 又三因, 四而一為池積.
[3.5] 今為方田積, 既以展起, 則此池積亦須展起. 故又用 1 步九分六釐乘
之, 得一步四分七釐, 亦為一個展起底圓池積也. 以一步九分六釐乘
之者, 蓋為分母十四. 以自之得一步九分六釐也.

[3.6] 以池積減田積餘 ⊟ [9]為一段如[10]積, 寄左.
[3.7] 然後列真積, 一萬一千三百二十八步, 亦用分母冪, 一步九分六釐, 乘
之得二萬二千二百零二步八分八釐.

---

[8]面徑 is in WYG and WJG
[9]47 instead of 0.47 in last line in WYG and WJG.
[10]虛 instead of 如 in WJG and WYG.

[3.8] 與左相消得 ⬚. [11] 平方開之得六十四步為內池徑也. 倍至步, 加池徑, 身外除四, 見方面也.

[3.9] 一法求所展池積. 以徑自之了, 更不須三因, 四除及, 以一步九分六釐乘之. 只於徑冪上以一步四分七釐. 乘之, 便為所展之池積也。

[3.10] 依條段求之. 展積內減四段至步冪, 餘為實. 四之至步為從. 四分七釐益隅.

[3.11] 義曰: 凡言展積, 者是於正積上, 以一步九分六釐乘起之數. 元法本是方面上寄一步四分. 分母自乘過, 於每步上, 得一步九分六釐. 故今命之為展起之數也. 諸變斜為方面[12]者皆準. 此所展之池積是, 於一步圓積上, 展出九分六釐.

[3.12] 若以池徑上取斜為外圓徑, 則一步上, 止生得四分七釐也. 故以四分七釐為虛常法.

Translation.

[3.1] Let us suppose there is one square piece of field, inside which there is a circular pond. Outside [the pond] occupied by water, one counts eleven thousand three hundred twenty-eight *bu* of land. It is said only that [the distances] from the angle of the outer square obliquely reaching the edge of the inner pond are fifty two *bu* each. One asks how much are of the inner diameter and the outer side each.

The answer: the side of the outer field is one hundred and twenty *bu*. The diameter of the inner pond is sixty four *bu*.

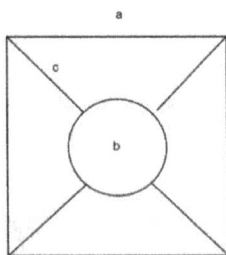

Fig. 9.6   a: square field; b: circular pond; c: fifty two *bu*.

[3.2] The method: Set up one Celestial Source as the diameter of the inner pond. Adding twice the *reaching bu* yields $\begin{smallmatrix}1\ 0\ 4\ tai\\1\end{smallmatrix}$ as a diagonal of the square.

---

[11]47 instead of 0.47 in last line in WJG and WYG
[12]面 is not in WYG.

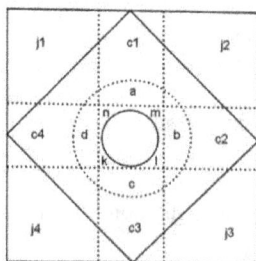

Fig. 9.7　j1–4: 減; c1–4: 從; abcd: 四分七釐; klmn: 二分五釐.[14]

$$
\begin{array}{c}
1\ 0\ 8\ 1\ 6\ \textit{tai} \\
2\ 0\ 8 \\
1
\end{array}
$$

[3.3] Augmenting this by self-multiplying yields as the square of the diagonal of the square, which is sent to the top.

From the diagonal of the square in the text above, one must remove out of the body its four [tenth]. Now, one does not carry out this diminution; one sends instead one bu four fen to make the denominator. Now, the square of the diagonal of the square is then the transformation of the diagonal into the side of a square. Besides, by self-multiplying this quantity, one obtains the expansion of this quantity.[15]

[3.4] One sets up further the Celestial Source as the diameter of the pond. This times itself, and increased further by three, then divided by four makes the area of the pond.

---

[14]Diagram is slightly different in WYG, the outer circle is passing through the corner of the inside square. See [Pollet (2014)] for the meaning of the difference between the two editions of diagrams.

[15]Let a be the side of the square and d, the diagonal. The purpose stated in the beginning of the problem is to find a. The first polynomial gives the diagonal. To find the side of the square from its diagonal, one just has to reduce the diagonal by $\sqrt{2}$. Here $\sqrt{2} = 1.4$. '*To diminish the diagonal by its four [tenth]*' is, in modern terms: $d - 1.4d = a$ or $\frac{d}{1.4} = a$. Instead of reducing the diagonal, Li Ye proposes another procedure. One keeps the diagonal to make an expanded square whose side is this diagonal and works with the values of this new square. 1.4 will be placed as denominator to reduce thereafter all the values. In the case of area, that is to use $1.4 \times 1.4 = 1.96$, which is '*the expansion of this quantity*'.

[3.5] Now, if one has expanded the area of the square field, then the area of the pond also needs to be expanded. That is why one uses further one *bu* nine *fen* six *li* to multiply this, what yields one *bu* four *fen* seven *li*, which also makes one expanded area of the inside circular pond.

One multiplies by one bu nine fen six li, because one takes fourteen as the denominator,[16] *which, self-multiplied, yields one bu nine fen six li.*

[3.6] Subtracting the area of the pond from the area of the field remains

```
1 0 8   1 6        tai
    2   0 8                    as one piece of the equal area which is sent to
      − 0 . 4 7
```

the left.

[3.7] After, one places the real area, eleven thousand three hundred and twenty eight *bu*. One uses also the square of the denominator, one *bu* nine *fen* six *li*, to multiply this.

It yields twenty two thousand two hundred and two *bu* eight *fen* eight *li*. With what is on the left, eliminating from one another

```
          −1 1 3   8 6 . 8 8 tai
yields         2   0 8
                 − 0 . 4 7
```

Opening the square of this yields sixty four *bu* as the diameter of the inside pond. One adds twice the *reaching bu* to the diameter of the pond, and reduces out of the body its four [tenth], there appears the side of the square.

[3.9] One looks for the expanded area of the pond according to another method. Once the diameter is self-multiplied, it is not necessary to increase by three, to divide by four, and to multiply this by one *bu* nine *fen* six *li*. One just directly multiplies the square of the diameter by one *bu* four *fen* seven *li*,

Commentary: this is to multiply the quantity of one bu nine fen six li by three and to divide it by four.

what makes the expanded area of the pond[17].

[3.10] One looks for this according to the section of pieces [of areas]. From the expanded area, one subtracts four pieces of the square of the *reaching bu* of the diagonal, what remains makes the dividend. Four times the *reaching bu* makes the adjunct. Four *fen* seven *li* is the augmented corner.

[3.11] The meaning: Each time one talks about expanded areas, one

---

[16]分母十四, *fen mu shi si*. The denominator is 1.4, which self-multiplied gives 1.96. The denominator was moved of one place on the counting board what makes a multiplication by 10.

[17]Let x be the diameter of the circular pond. In the previous method, the expanded area of the pond was: $\frac{3}{4}x^2) \times 1.96 = 1.47x^2$. The alternative method suggests to combine the coefficients in order to have only one coefficient, that is to directly multiply x^2 by 1.47, the latter being the result of $\frac{3}{4} \times 1.96$.

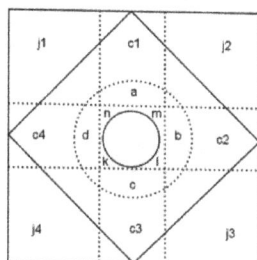

Fig. 9.8 j1–4: subtract; c1–4: adjunct; abcd: four *fen* seven *li*; klmn: two *fen* five *li*.

means a quantity which emerges from a multiplication of a real area by one *bu* nine *fen* six *li*. The original method basically consisted in sending one *bu* four *fen* on the side of the square. Once the denominator is self-multiplied, for each *bu*, it yields one *bu* nine *fen* six *li*. Therefore, now, one names this 'to make the expanded value'. Every transformations of a diagonal into a side of the square follow the same standard. The area of the pond which is expanded is that for each *bu* of the circular area, one expands by nine *fen* six *li*.

[3.12] If, on the diameter of the pond, one takes the diagonal to make the diameter of the outer circle, hence, on one *bu*, it produces only four *fen* seven *li*. Therefore, four *fen* seven *li* makes the empty constant divisor.

Description.

[3.1] Let a be the distance from the angle of the square field, to the circle, 52 *bu* ; let A be the area of the square field (S) less the area of the circular pond (C), 11328 *bu*; and x be the diameter of the pond.

The procedure of the Celestial Source:

[3.2] Diagonal of the square $= 2a + x = 104 + x$.

[3.3] Expanded area of S: $S \times (1.4)^2 = (2a + x)^2 = 4a^2 + 4ax + x^2 = 10816 + 208x + x^2$.

[3.4] $C = \frac{3}{4}x^2 = 0.75x^2$, since $\pi = 3$.

[3.5] Expanded area of C: $1.96C = \frac{3}{4}x^2 \times (1.4)^2 = 1.47x^2$.

[3.6] $1.96S - 1.96C = 4a^2 + 4ax + x^2 - (\frac{3}{4}x^2 \times (1.4)^2) = A \times (1.4)^2$.

[3.7] $= 10816 + 208x + x^2 - 1.47x^2 = 10816 + 208x - 0.47x^2 = 11328 \times 1.96 = 22202.88$ *bu*.

[3.8] The equation: $(4a^2 - 1.96A) + 4ax + x^2 - 1.47x^2 = -11386.88 + 208x - 0.47x^2 = 0$.

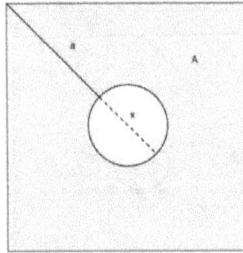

Fig. 9.9

The procedure by Section of Pieces [of Areas]:

$1.96A - 4a^2$	dividend
$4ax$	adjunct
$-0.47x^2$	augmented corner

A piece of the diagonal, a, is given with the area represented in pink in [Fig. 9.10], and one looks for the diameter of the interior circle. Li Ye, expressing the diagonal of the square in term of the unknown $(2a + x)$, transforms the diagonal into the side of a square. Instead of dividing by $\sqrt{2}$ (value of diagonal = side of the square with denominator $\sqrt{2}$), he multiplies every length by 1.4. [3.3] reads '*The diagonal of the square (···) corresponds to an outer body diminished by four. Now, one does not carry out this diminution; one places instead one bu four fen as denominator*'. In the case of area, that means to multiply by 1.96, and '*one names this 'to make the expanded value*' [3.11]. Once, one has transformed the side of the square into a diagonal by using $\sqrt{2}$, and expanded all the dimensions [Fig. 9.11], one can, hence, proceeds exactly like in the Problem 1 [Fig. 9.12, see also Problem 1]. First, one removes the four squares of side a which are in the corners. It remains four rectangles of length a and width x. Thus, $4ax$ makes the adjunct divisor. Then to find the coefficient of the term in x^2, one has to consider the central square area of side x. From this area, one removes the expanded area of the circle. That is $-1.47x^2 + x^2$. To find the expanded area of the circle, one looks for the diagonal of the central square and uses it as a diameter. [3.12] '*on the diameter of the pond, one takes the diagonal [Fig. 9.13] to make the diameter of the outer circle*' [Fig. 9.14]. The difference between these two areas is the constant divisor: $-0.47x^2$. '*hence, on each bu, it produces only four fen seven li. Therefore, four fen seven li is the empty constant divisor*' [Fig. 9.15]. So the equation reads: $1.96A - 4a^2 = 4ax - 0.47x^2$.

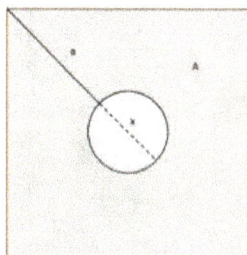

Fig. 9.10

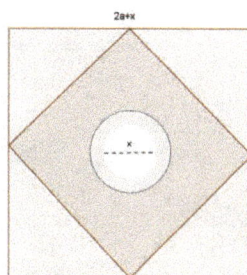

Fig. 9.11

Problem 36

第三十六問

[36.1] 今有圓田一段, 中心有直池. 水占之外計地六千步. 只云從內池四
角斜至田楞各一十七步半. 其內池長闊共相和得八十五步.
問三事各多少.
荅曰: 外田徑一百步. 池長六十步. 闊二十五步.

[36.2] 法曰: 立天元一為內池斜. 加入倍至步, 三十五, 得 ⊞ 為外圓徑.

[36.3] 以自之, 又三之得 ⊞ 為四段圓積也.

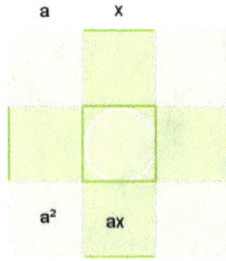

Fig. 9.12

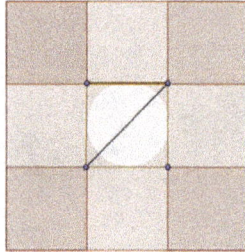

Fig. 9.13

[36.4] 內減四之見積, 二萬四千步, 得下 ⬚ 為四個池積, 寄左. 乃置內池和, 八十五步.

[36.5] 以自之得 ⬚ 為四積一較冪, 於頭.

[36.6] 再立天元內池斜. 以自之得 ⬚ 為二池積一較冪.

[36.7] 以減於頭位得 ⬚ 為二池積也.

[36.8] 又倍之得 ⬚ 亦為四池積.

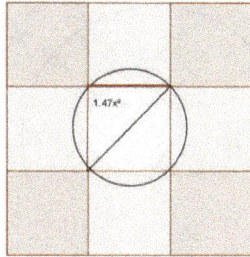

Fig. 9.14

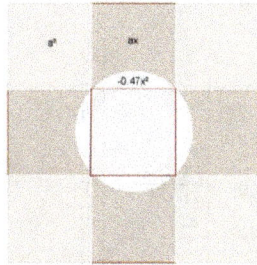

Fig. 9.15

[36.9] 與左相消得 ⌐⌐⌐. 平方開得六十五步為內池斜. 加倍至步, 即圓徑也. 徑自之, 又三之四而一, 內減去田積餘實. 以和步為從一. 虛隅. 開平方見闊也.

[36.10] 依條段求之. 四之積步內加兩段和步冪, 卻減十二段至步冪為實. 十二之至步為從. 五步常法.

[36.11] 義曰: 所加兩個和冪, 該八積二較冪數. 內元有四虛池, 外有四積二較冪.

[36.12] 其實只是添了兩個池斜冪也. 於四圓積內除從步占. 外元有三個方. 今又加入兩[18]個池斜冪, 共得五步. 故五為常法.

Translation.

[36.1] Let us suppose there is one circular piece of field in the middle of which there is a rectangular pond. Outside the [area] occupied by

---

[18] 二 instead of 兩 in WJG.

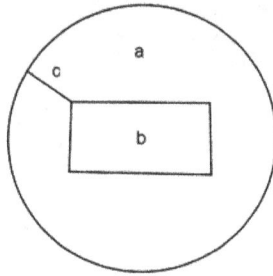

Fig. 9.16   a: 圓田; b: 直池; c: 八十七步半.

Fig. 9.17   a1–2: 加; b: 元有; c1–12: 從; J1–12: 減.

water, one counts six thousand *bu* of land. It is said only that the diagonals from the four angles of the inner pond reaching the edge of the field are seventeen and a half *bu* each. Mutually summed up together, the length and the width of the inner pond yields eighty-five *bu*.

One asks how much these three things are.

The answer: the diameter of the outer field is one hundred *bu*. The length of the pond is sixty *bu*. The width is twenty-five *bu*.

[36.2] The method: Set up one Celestial Source as the diagonal of the inner pond. Adding twice the *reaching bu*, thirty five, yields $\begin{smallmatrix} 3 & 5 \ tai \\ & 1 \end{smallmatrix}$ as the diameter of the outer circle.

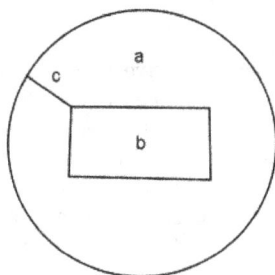

Fig. 9.18   a: circular field; b: rectangular pond; c: seventeen and a half *bu*.

[36.3] This times itself, and further by three yields $\begin{smallmatrix} 3\ 6\ 7\ 5 \\ 2\ 1\ 0 \\ 3 \end{smallmatrix}$ [19] as four

pieces of the circular area.

[36.4] From this, one subtract four times the real area, twenty four

thousand *bu*, it yields the following: $\begin{smallmatrix} -2\ 0\ 3\ 2\ 5 \\ 2\ 1\ 0 \\ 3 \end{smallmatrix}$ as four areas of

the pond, which are placed on the left.[20] Then, put down [the *bu* of] *the sum* of the inner pond, eighty five *bu*.

[36.5] This times itself yields 7225 *tai*[21] as four areas [of the pond] and one square of *the difference*, which is sent to the top.

[36.6] Set up again the Celestial Source, the diagonal of the inner pond.

This times itself yields $\begin{smallmatrix} 0\ yuan \\ 1 \end{smallmatrix}$ as two areas of the pond and one

square of *the difference*.

[36.7] Subtracting this from what is on the top position yields $\begin{smallmatrix} 7\ 2\ 2\ 5 \\ 0 \\ -1 \end{smallmatrix}$

as two areas of the pond.

[36.8] Double this further yields $\begin{smallmatrix} 1\ 4\ 4\ 5\ 0 \\ 0 \\ -2 \end{smallmatrix}$ which are also four areas of

the pond.

[36.9] With what is on left, eliminating them from one another yields

---

[19]The character *tai* 太 is not written any further in this problem.
[20]The left position is mentioned first, then secondly the top position.
[21]The writing is in the style of counting rods.

−3 4 7 7 5
2 1 0
5

Open the square yields sixty-five *bu* as the diagonal of the inside pond. Adding twice *the reaching bu* gives the diameter of the circle. [Multiply] the diameter by itself, then further by three and divide it by four. From this, one subtracts the area of the field, it remains the dividend. The *bu of the sum* makes the adjunct. One is the empty corner. Open the square, there appears the width[22].

[36.10] One looks for this according to the section of pieces [of areas]. Four times the *bu* of the area are added to two pieces of the square of the *bu of the sum*. Conversely, one subtracts twelve pieces of the square of *the reaching bu* to make the dividend. Twelve times *the reaching bu* makes the adjunct. Five *bu* is the constant divisor.

[36.11] The meaning: the two squares of *the sum* that are added [to the real area] equal the quantity of eight areas [of the pond] and two squares of the comparison. Inside of the original [area], there are four empty ponds; outside there are four areas [of the pond] and two squares of *the difference*.

Fig. 9.19   a1–2: To add; b: 'the original [area] has'; c1–12: adjunct; J1–12: to subtract.

[36.12] [To make] this dividend, one only has to complement with two squares of the diagonal of the pond. Inside four circular areas, one removes [the part] that is filled [in between] the *bu* of the adjunct, outside of the original [area], there are three squares. Now, one adds further two squares of the diagonal of the pond together, it yields five *bu*. Therefore, five makes the constant divisor.

---

[22]The length that was asked is not given here.

Description.

[36.1] Let a be the distance going from the circle to the angle of the rectangular pond, 17.5 bu ; let b be the length added to the width, 85 bu; and d, their difference. Let A be the area of the circular field (C) less the area of the rectangular pond (R), 6000 bu; and x be the diagonal of the pond.

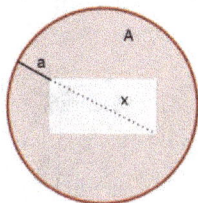

Fig. 9.20

The Procedure of the Celestial Source:

[36.2] To express R according to A:
The diameter $= 2a + x = 35 + x$.

[36.3] $3\times$ the square of the diameter $= 3(2a+x)^2 = 12a^2 + 12ax + 3x^2 = 3675 + 210x + 3x^2 = 4C$.

[36.4] $4C - 4A = 12a^2 + 12ax + 3x^2 - 4A = -20325 + 210x + 3x^2 = 4R$.

[36.5] To express R according to b:
The square of the length added to the width $= b^2 = 4R + d^2 = 7225$.

[36.6] The square of the diagonal $= x^2 = 2R + d^2$.

[36.7] $b^2 - x^2 = 7225 - x^2 = 2R$.

[36.8] $2 \times 2R = 2b^2 - 2x^2 = 14450 - 2x^2 = 4R$.

[36.9] The equation: $(4C - 4A) - 4R = 12a^2 + 12ax + 3x^2 - 4A - (2b^2 - 2x^2) = -4A - 2b^2 + 12a^2 + 12ax + 5x^2 = -34775 + 210x + 5x^2 = 0$.

Problem 14

第十四問

[14.1] 今有圓田一段, 內有方池. 水占之外計地三百四十七步. 只云從田外楞通內池斜三十五步半.
問外圓內方各多少.
荅曰: 外圓徑三十六步. 內方面二十五步.

[14.2] 法曰: 立天元一為內方面. 加四得 [步] [23] 為方斜.

---

[23] 步 is not in WJG and WYG.

[14.3] 以減倍通步得 |太／長| 為外圓徑.

[14.4] 以自增乘得 |▦| 為外田徑冪也.

[14.5] 以三之得 |▦| 為四段圓田積, 於頭.

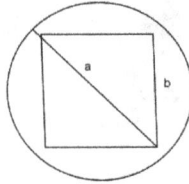

Fig. 9.21   a: 三十五步半; b: 方面二十五步.

[14.6] 再立天元內方面. 以自之, 又就分四之得 |元／IIII| 為四段方池.

[14.7] 以減頭位得 |▦| 為四段如積, 寄左. 然後列四之見積, 一千三百八十八步.

[14.8] 與左相消得 |▦|. 開平方得二十五步為內方面也. 方面加四, 減於倍通步得圓徑也.

[14.9] 依條段求之. 十二段通步冪內減四之田積為實. 十二之通步加四為益從. 一步八分八釐常法.

[14.10] 義曰: 此式原²⁴係虛從. 今以虛隅命之.

[14.11] 四段圓田減積時, 剩下四段方池²⁵. 於從步內, 用訖三個. 外猶剩一個. 卻於二步八分八釐虛數內, 補了一步, 外虛一步八分八釐. 故以之為法.

[14.12] [從負, 隅正, 或從正, 隅負, 其實皆同故因此廉從以別之。]

[14.13] 舊術曰: 倍通步. 自乘三之為頭位. 四因田積. 減頭位, 餘為實. 又十二通步. 加四為從法. 廉常置一步八分八釐. 減從. 開方. [新舊廉從不同. 開時, 則同. 故兩存之.]

---

²⁴元 instead of 原 in WJG and WYG.

²⁵田 instead of 池 in WJG.

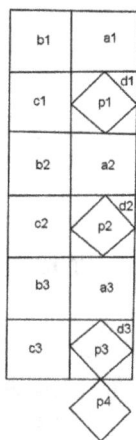

Fig. 9.22　a1–3: 連下方面二之從; b1–3: 減徑方, 餘一方池又三分之一; c1–3: 連右方面二之從; d1–3: 九分六釐; p1–4: 池.

Translation.

[14.1] Let us suppose there is one piece of circular field, inside of which there is a square pond. Outside the [area] occupied by water, one counts three hundred forty-seven *bu* of land. It is said only that [the distance] from the outer edge of the field *going through* the diagonal of the inner pond is thirty-five *bu* and a half.

One asks how much are the diameter of the outer circle and the sides of the inner square each.

The answer: the diameter of the outer circle is thirty-six *bu*; the side of the inner square is twenty-five *bu*.

[14.2] The method: Set up one Celestial Source as the side of the inner square. Augmenting it by four [tenths] yields $\begin{smallmatrix} 1 \, . \, 4 \; yuan \\ bu \, . \end{smallmatrix}$ as the diagonal of the square.

[14.3] Subtracting this from twice the *bu* *going through* yields $\begin{smallmatrix} 7\,1 \quad tai \\ -\,1\,.\,4 \end{smallmatrix}$ as the diameter of the outer circle.

[14.4] Augmenting this by self-multiplying yields $\begin{smallmatrix} 5\,0\,4\,1 \\ -\,1\,9\,8\,.\,8 \\ 1\,.\,9\,6 \end{smallmatrix}$ [26] as the square of the diameter of the outer field.

---

[26] The character *tai* is not written in all the following polynomials.

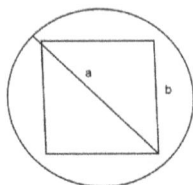

Fig. 9.23   a: thirty five *bu* and a half; b: side of the square, twenty-five *bu*.

                                    1 5 1 2 3
[14.5] Triple this yields   − 5 9 6 . 4   as four pieces of the area of the
                                        5 . 8 8
circular field, which is sent to the top.

[14.6] Set up again the Celestial Source, the side of the inner square.
This times itself and, with the help of the part, to quadruple this
yields $\frac{0\ yuan}{4}$ as four pieces of the square pond.

[14.7] Subtracting from what is on the top position yields
        1 5 1 2 3
yields   − 5 9 6 . 4   as four pieces of the equal area, which is sent to
              1 . 8 8
the left.
After, place the real area. [Place] four times, one thousand three
hundred eighty-eight *bu*.

[14.8] With what is on the left, eliminating them from one another
        1 3 7 3 5
yields   − 5 9 6 . 4
              1 . 8 8
Opening the square yields twenty-five *bu* as the side of the inner
square. Augment the side of the square by four [tenths], subtract
this from twice the *bu going through*; it yields the diameter of the
circle.

[14.9] One looks for this according to the section of pieces [of areas].
From twelve pieces of the square of the *bu going through*, one subtracts four times the area of the field to make the dividend. Twelve
times the *bu going through* augmented by four [tenths] makes the
augmented adjunct. One *bu* eight *fen* eight *li* is the constant divisor.

[14.10]The meaning: the pattern[27] originally has an empty adjunct.
Now, I recommend the use of an empty corner.

[14.11] When one subtracts four pieces of the circular field from the area,

---
[27]*shi* 式.

b1 a1

c1 p1 d1

b2 a2

c2 p2 d2

b3 a3

c3 p3 d3

p4

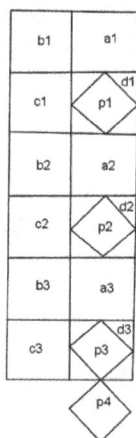

Fig. 9.24 a1–3: together with the side of the square below, two times [the *bu through* makes] the adjunct; b1–3: one subtracts the square of the diameter, but it remains one square pond and one third; c1–3: together with the side of the square on right, two times [the *bu through* makes] the adjunct; d1–3: nine *fen* six *li*; p1–4: pond.

there remains the following four pieces of square ponds. Inside the *bu* of the adjunct, once one used the three [ponds], outside it still remains one [pond]. Conversely, on each empty quantity of two *bu* eight *fen* eight *li*, once one compensated one *bu*, outside there are one empty *bu* eight *fen* eight *li*. Therefore, with this one makes the [constant] divisor.

[14.12] *[Whatever there is] a negative adjunct with a positive corner or a positive adjunct with negative corner, the dividend amounts always the same. That's why one uses the edge and the adjunct to differentiate.*

[14.13] The old procedure: twice the *bu going through*, self-multiply them and multiply by three to make what is at the top position. Multiply by four the area of the field, and subtract it from what is sent to the top position. The remainder makes the dividend. [Put] further twelve times the *bu going through*; augment it by four to make the adjunct divisor. The edge-constant [divisor] is one *bu* eight *fen* eight *li*. Subtract the adjunct and open the square.
In the new and the old (procedures), the edge (divisor) and the adjunct (divisor) are not the same. But when one opening (the square), then (the process) is the same. Therefore, one has to preserve these two (procedures).

Description.

[14.1] Let a be the distance leaving from the circle and going along the diagonal of the square, 35.5 *bu*; let A be the area of the circular field (C) less the area of the square pond (S), 347 *bu*; and x be the side of the pond.

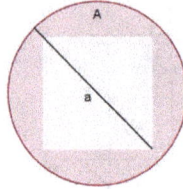

Fig. 9.25

The procedure of the Celestial Source:

[14.2] Diagonal of the square $= 1.4x$;

[14.3] Diameter $= 2a - 1.4x = 71 - 1.4x$;

[14.4] Square of the diameter $= (2a - 1.4x)^2 = 4a^2 - 5.6ax + 1.96x^2 = 5041 - 198.5x + 1.96x^2$;

[14.5] $4C = 3(4a^2 - 5.6ax + 1.96x^2) = 12a^2 - 16.8ax + 5.88x^2 = 15123 - 596.4x + 5.88x^2$;

[14.6] $4S = 4x^2$;

[14.7] $4C - 4S = 12a^2 - 16.8ax + 1.88x^2 = 4A$;

    $= 15123 - 596.4x + 1.88x^2 = 1388$;

[14.8] The equation: $12a^2 - 4A - 16.8ax + 1.88x^2 = 13735 - 596.4x + 1.88x^2 = 0$.

The procedure by Section of Pieces of Areas:

$12a^2 = 4A + 2 \times 6 \times 1.4ax - 3 \times 1.96x^2 + 4x^2$;

$12a^2 - 4A = 2 \times 6 \times 1.4ax - 5.88x^2 + 4x^2$;

$12a^2 - 4A = 2 \times 6 \times 1.4ax - 2.88x^2 + x^2$;

The equation: $12a^2 - 4A = 2 \times 6 \times 1.4ax - 1.88x^2$.

To solve this problem, we must have in mind the two following rules which are given in some previous problems:

(1) 4 areas of a circle make 3 areas of squares whose side is the diameter, and one considers the approximate value of $\pi = 3$.

(2) To find the diagonal of a square, one has to multiply the side by $\sqrt{2}$, which is approximated in the *Development of Pieces [of Areas]* as $\sqrt{2} = 1.4$. This operation is named *jia si* 加四 by Li Ye, which we translate as 'augment by four [tenths]'. (See translation

of Problem3).

The first sentence of the Section of Pieces [of Areas] gives operations leading to coefficients. Using the symbolic transcription of the statement of the problem, we have the following elements:

$12a^2 - 4A$	dividend
$12 \times 1.4a$	augmented adjunct
1.88	constant divisor

This problem presents the same difficulties as Problem 18. If one reads yi $cong$, as 'negative adjunct' we should transcribe the equation as: $12a^2 - 4A = -12 \times 1.4ax + 1.88x^2$. And this equation is not equal to the one presented in the procedure of the Celestial Source. We should transcribe either $12a^2 - 4A - 12 \times 1.4ax + 1.88x^2 = 0$ or $12a^2 - 4A = 12 \times 1.4ax - 1.88x^2$. The second transcription is more convincing for the following reason:

Since four areas of the circle are equal to three squares whose side is the diameter, one starts with representing three squares whose sides are $2a$. One does not know the diameter d. So, one will use a to construct the squares because it is the only available constant and $2a$ because this segment allows one to express the diameter on the basis of the diagonal of the square pond, the latter being the expanded side of the square pond whose side is the unknown. That is $2a = d + 1.4x$. That means the diameter is: $d = 2a - 1.4x$. Therefore, one has 3 squares corresponding to $12a^2$, whose total area is a known constant. See [Fig. 9.26]. We also know from the statement that $4C = 4A + 4S$ and, from previous problems, that $3d^2 = 4C$. From this area made of squares whose side is $2a$, one removes $3d^2$. See [Fig. 9.27].

After the subtraction, the remaining area can be translated into an area composed of 6 rectangles whose length is $2a$ and width is $1.4x$, the unknown that one is looking for. These rectangles represent the adjunct divisor. But all these rectangles are stacked on one square area: $(1.4x)^2$. See [Fig. 9.28]. These three square areas are in excess and must be removed. In [Fig. 9.29], the green part represents $6 \times 2a \times 1.4x - 3 \times (1.4x)^2$ and this is equal to $12a^2 - 3d^2$. Therefore, one has $12a^2 - 3d^2 = 6 \times 2a \times 1.4x - 5.88x^2$.

But the removal of $3d^2$ is equivalent to the removal of $4C$. $4C = 4S + 4A$. $4A$ is a constant given in the statement and $4S = 4x^2$. Thus $12a^2 - 3d^2 = (12a^2 - 4A) - 4x^2$. This is what Li Ye means by [14.11] '*When one subtracts four pieces of the circular field from the area (i.e. $12a^2 - 4A$), there remains the following four pieces of the square ponds*' in the 'meaning'. In fact, in $12a^2 - 3d^2 = 6 \times 2a \times 1.4x - 5.88x^2$, one also removed four squares of side x. That is, one removed 'too much space'. One lost $4S$, and had to compensate. To compensate for this loss, Li Ye proceeds in two steps. First, add three squares whose side is

Fig. 9.26

Fig. 9.27

unknown to each of the bottom right corners. In [Fig. 9.30], the green part represents $6 \times 2a \times 1.4x - 3 \times (1.4x)^2 + 3x^2$. Next compensate again by adding another extra square pond, which will be outside at the bottom. This is why Li Ye says [14.11]: '*Inside the bu of the adjunct, once one used the three [ponds], outside it still remains one [pond]*'. There was thus: $-3 \times (1.4x)^2 + 3x^2 = -2.88x^2$; and now, to this, one compensates further $1x^2$. That is $-2.88x^2 + x^2$. Li Ye expresses this in the following way [14.11]: '*on each empty quantity of two bu eight fen eight li, once one*

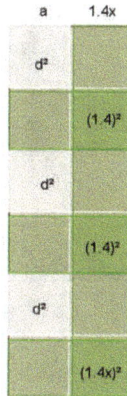

Fig. 9.28

Fig. 9.29

compensated one bu, outside there are one empty bu eight fen eight li' and therefore one finds $-1.88x^2$ as a 'constant divisor'. The final diagram [Fig. 9.32] represents $12a^2 = 4A + 6 \times 2a \times 1.4x - 1.88x^2$. Which can also be read as $12a^2 - 4A = 6 \times 2a \times 1.4x - 1.88x^2$.

At the end of the procedure above, one can transcribe the equation as: $12a^2 - 4A = 12a \times 1.4x - 1.88x^2$. The dividend $(12a^2 - 4A)$ is thus 'positive', as is the adjunct $(6 \times 2a \times 1.4x)$, while the corner/constant divisor $(-1.88x^2)$ is 'negative.' This transcription is different from the

Fig. 9.30

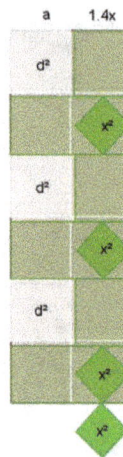

Fig. 9.31

first one suggested at the beginning where *yi* is transcribed by a negative coefficient: $12a^2 - 4A = -12a \times 1.4x + 1.88x^2$.

This is where the recommendation by Li Ye will make sense [14.10]: '*the pattern originally had an empty adjunct. Now, I recommend the use of an empty the corner*'. After one compensates the loss of area by adding back four squares of side x. In fact, Li Ye, in his recommendation, proposed to empty the four triangular corners of each of the squares (of side $1.4x$) which are inside the rectangular area representing the adjunct directly. His recommendation has to be read literally: one should now empty some corners, instead of emptying the adjunct. I represented the corners that have to be emptied in blue in [Fig. 9.32]. It means one does not have to remove the three extra squares of the adjunct and to compensate the loss of area any more. One can spare the steps of the procedure that are represented on [Fig. 9.29] and [Fig. 9.30]. It is a question of economy in the procedure. Yet, the operations present in [Fig. 9.29] and [Fig. 9.30] makes that the Problem 14 mirroring Problem 4. Exactly the same proposition is made for Problem 18 in exactly the same situation.

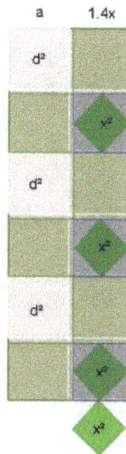

Fig. 9.32

The old procedure: [14.13]

Dividend $=3(2a^2) - 4A$;
Adjunct divisor $=12 \times 1.4a$;
Edge constant divisor $=1.88$;
The equation: $3(a^2) - 4A = 12 \times 1.4ax - 1.88x^2$.

CONCLUSION

Li Ye sought to make the content of an ancient book, the *Collection Augmenting the Ancient [knowledge]* (eleventh century) more accessible to readers by illustrating it with diagrams. A systematic study of the diagrams shows that one of the most important features of the *Development of Pieces [of Areas]* is the practice of transformation of figures. The heart of the book rests on non-discursive practices: drawing and visualising the transformation of figures and manipulating counting rods on a support. The topic is reflected in the title, *Yigu yanduan*, now translated as 'Development of Pieces [of Areas] [according to] [the collection] Augmenting the Ancient [Knowledge]'. That is, the procedure of development by the Section of Pieces of Areas presented in the treatise titled the *Collection Augmenting the Ancient [knowledge]* (augmented with diagrams).

Li Ye wanted to transmit this treatise because he found it as remarkable as the commentary by Liu Hui to the famous classic, the *Nine Chapters*. In the *Development of Pieces [of Areas]*, the problems are ordered according to analogy using reduction and iteration. The analogical reasoning forms an important feature in reading both treatises. The *Development of Pieces [of Areas]* shares some practices with the classic text and its commentary. These practices involve a specific way of transforming diagrams, a way of ordering problems, a way of using the analogy for demonstration and a way of giving a geometric account of quadratic equations and the extraction of square roots. However, if these practices have common features, they also have differences: there are mental 'manipulations' of figures in the *Development of Pieces [of Areas]* instead of the material manipulations of the Han Dynasty; there is a multiplicity of problems as examples in the *Development of Pieces [of Areas]*, whereas a unique example, used as a model, is sufficient in the *Nine Chapters*. All of these similarities and differences in practice demonstrate the broken continuity and lacunae in algebraic practices in China from the Han to the Song-Yuan period.

But what is the real aim of the treatise? Is Li Ye's purpose the same as that of the author of the *Collection Augmenting the Ancient [knowledge]*? According to the extant materials, Liu Yi was the first to manage to solve arbitrary higher-degree equations using the method of iterated multiplication. Crossley translates Yang Hui's preface of the *Fast Methods of Multiplication and Division* as '*Master Liu of Zhongshan [···] introduced the corollary to the method of extracting square root independent of positive and negative, which had never been heard before*' [Li and Du (1987), 128]. The work of Liu Yi is concerned with expanding the discipline of 'opening the square', that is, of finding solutions to equations of the form $x^2 + ax = A$. This algorithm is also used in a more general situation where products can be added or subtracted, which means that one could consider cases with negative coefficients, even if those are not explored yet [Horiuchi (2000), 244]. This expansion of the procedure is legitimated by placing it within a geometrical support which is later presented in the Section of Pieces [of Areas] procedure by Yang Hui. Thus, historians[28] consider Liu Yi to have introduced equations with negative coefficients. The *Development of Pieces [of Areas]* shows an evolution of diagrammatic expression of quadratic equation from positive to negative coefficient evolution, which was already achieved at Li Ye's time.

Liu Yi is famous for considering negative coefficients and geometrical representations of equations including 'adjunct' rectangles and 'added corners.' This innovation framed the quadratic equation in terms of the geometrical justification of the extraction of square roots. Li Ye gives the only example of the specific use of the character *xu*, 'empty,' in this context. Chapter 5 showed the evolution of diagram and numbers of 'adjunct' rectangles from an old to a new procedure. The *Development of Pieces [of Areas]* reveals investigations of geometrical interpretations of negative coefficients and multiple 'adjunct.' It is a work in a specific mathematical discipline, the geometric construction of algebraic equations through the combination of several adjunct rectangles. In fact, the work transmitted by Li Ye could be the oldest evidence of the exploration of what we call equations, and the first evidence of polynomial computation in China. It also might be that the *Collection Augmenting the Ancient [knowledge]* was a sophisticated exploration of geometrical demonstrations of algorithms

---

[28][Li and Du (1987)]; [Te (1990)]; [Lam (1977)]; [Horiuchi (2000)].

for setting up quadratic equations with negative and positive coefficients, which were new mathematical objects in the eleventh century. The construction of negative coefficients plays a pivotal role in this evolution. It also is possible to distinguish several layers of composition that reflect several episodes in the meandering development of the quadratic equation. In other words, this analysis raises a philological problem pertaining to the question of textual transmission and the nature of authorship.

The most interesting feature of the *Development of Pieces [of Areas]* is the order of its sixty-four problems. They are arranged according to a sophisticated architecture combining sets of types of geometrical figures, sets of mathematical objects, sets of operations of transformation of diagrams and sets of data given in the statement of problem. Problems are arranged according to their analogical resemblance. They are reduced to one another and arranged according alternation and iteration. They refer to each other tacitly and this creates a web expressing generality. This order appears only in the geometrical procedure, when diagrams are individually reproduced. It is not possible to modify the order of the problems. The creator of this web was conducting explorations on Changes, and mathematics was a WAY to do so.

Appendix A

Supplements

A.1 Tables

All equations are expressed according to the data given in the statement of the problem and named here by capital or lower case letters.

A: capital letters express areas given in the statement of the problem.

a: lower case letters express the distances given in the statement of the problem.

Div:-: Dividend is negative. In all other cases, the dividend is positive.

Table A.1: Equations in the *Development of Pieces [of Areas]*.

Problem	Celestial Source	Sections of Areas	Old Procedure
1	$A - 4a^2 - 4ax - 0.25x^2 = 0$	$A - 4a^2 = 4ax + 0.25x^2$	
2	$A - 4a^2 + 4ax - 0.25x^2 = 0$ Div:-	$4a^2 - A = 4ax - 0.25x^2$	
3	$(4a^2 - 1.96A) + 4ax - 0.47x^2 = 0$ Div:-	$1.96A - 4a^2 = 4ax - 0.47x^2$	
4	$4a^2 - 1.96A - 4ax - 0.47x^2 = 0$	$4a^2 - 1.96A = 4ax + 0.47x^2$	
5	$3a^2 - 48A + 6ax - x^2 = 0$ Div:-	$48A - 3a^2 = 6ax - x^2$	$3(16A - a^2) = 6ax - x^2$
6.1	$12A - 11x^2 = 0$	$12A = 11x^2$	$\frac{12A}{11} = x^2$

6.2	$9A - 8.25x^2 = 0$	$9A = 8.25x^2$	
6.3	$A - 8.25x^2 = 0$	$A = 8.25x^2$	$\frac{A}{8.25} = x^2$
7	$-12A - a^2 - 2ax + 11x^2 = 0$	$12A + a^2 = -2ax + 11x^2$	
8	$a^2 - 16A - 6ax - 3x^2 = 0$	$a^2 - 16A = 6ax + 3x^2$	$\frac{a^2-16A}{6} = ax + 0.5x^2$
9	$4a^2 - 81A - 20ax - 35.75x^2 = 0$	$4a^2 - 81A = 20ax + 35.75x^2$	$\frac{4a^2-81A}{10} = 2ax + 3.575x^2$
10	$100a^2 - 2209A - 500ax - 1031.75x^2 = 0$	$(10a)^2 - 2209A = 500ax + 1031.75x^2$	$3((10a)^2 - 2209A) = 1500ax + 3095.25x^2$
11.A	$12a^2 - 4A + 12ax - x^2 = 0$ Div:-	$4A - 12a^2 = 12ax - x^2$	
11.B	$5.88a^2 - 4 \times 1.96A + 11.76ax + 1.88x^2 = 0$ Div:-	$4 \times 1.96A - 3 \times 1.96a^2 = 1.96 \times 6ax + 1.88x^2$	
12	$12a^2 - 4A - 12ax - x^2 = 0$	$12a^2 - 4A = 12ax + x^2$	
13	$12a^2 - 4A + 16.8ax + 1.88x^2 = 0$ Div:-	$4A - 12a^2 = 12 \times 1.4ax + 1.88x^2$	$\frac{4a-3(2a)^2}{2} = 3(2a) \times 1.4x + 0.94x^2$
14	$12a^2 - 4A - 16.8ax + 1.88x^2 = 0$	$12a^2 - 4A = 2a \times 6 \times 1.4x - 1.88x^2$	$3(2a^2) - 4A = 12 \times 1.4ax - 1.88x^2$
15	$a^2 - 12A + 8ax + 4x^2 = 0$ Div:-	$12A - a^2 = 8ax + 4x^2$	$\frac{12A-a}{8} = ax + 0.5x^2$
16	$-16A + 11x^2 = 0$ Div:-	$12d^2 = 11x^2$	$\frac{16A}{11} = x$
17	$-16A - a^2 - 2ax + 11x^2 = 0$ Div:-	$16A + a^2 = -2ax + 11x^2$ Xu cong	

18	$a^2 - 12A - 8ax + 4x^2 = 0$	$a^2 - 12A = -8ax + 4x^2$ (incorrect reading) Corrected: $a^2 - 12A = 8ax - 4x^2$	$\frac{a^2-12A}{8} = ax - 0.5x^2$
19	$3a^2 - 4A - 6ax - x^2 = 0$	$3a^2 - 4A = 6ax + x^2$	$\frac{3(2c)^2 - 196A}{14} = 6cx + 3.5x^2 \;(c \neq a)$
20	$300a^2 - 4900A - 1980ax - 1633x^2 = 0$	$3(10a^2) - 4900A = 1980ax + 1633x^2$	
21	$3a^2 - (A + B + C) + 2x^2 = 0$ Div:-	$A + B + C - 3a^2 = 2x^2$	$\frac{A+B+C)-3a^2}{2} = x^2$
22	$a^2 - 1.96A + 2ax - 0.96x^2 = 0$ Div:-	$1.96A - a^2 = 2ax - 0.96x^2$	$A - (a/1.4)^2 = 2(a/1.4)x - 0.96x^2$
23	$14a^2 - 14A + 28ax + 25x^2 = 0$ Div:-	$14A - 14a^2 = 28ax + 25x^2$	
24	$A - a^2 + 2ax - 1.75x^2 = 0$ Div:-	$a^2 - A = 2ax - 1.75x^2$	
25	$11a^2 - 176A + 22ax + 25x^2 = 0$ Div:-	$176A - 11a^2 = 22ax + 25x^2$	
26	$48A - (3a^2 - 6ax + 7x^2) = 0$ Div:-	$3a^2 - 48A = 6ax - 7x^2$	
27	$11a^2 - 14A + 22ax + 25x^2 = 0$ Div:-	$14A - 11a^2 = 22ax + 25x^2$	
28	$14a^2 - 176A + 28ax + 25x^2 = 0$ Div:-	$176A - 14a^2 = 28ax + 25x^2$	

29	$48A - (4a^2 - 8ax + 7x^2)$ $\quad = \quad 0$ Div:-	$4a^2 - 48A = 8ax - 7x^2$	
30	$28A - (21a^2 - 42ax + 43x^2) = 0$ Div:-	$21a^2 - 28A =$ $42ax - 43x^2$	
31	$4a^2 - d^2 - 2A - 4ax - 0.5x^2 = 0$	$2a^2 - 2A - d^2 =$ $4ax + 0.5x^2$	
32	$-4A + a^2 + 2x^2 = 0$ Div:-	$4A - a^2 = 2x^2$	
33	$-4A - a^2 + b^2 + 2ax + 2x^2 = 0$ Div:-	$4A - b^2 + a^2 =$ $2ax + 2x^2$	$\frac{4A+a^2-b^2}{2} = ax + x^2$
34	$-4A + 8ax + x^2 = 0$ Div:-	$4A = 8ax + x^2$	
35	$-4A + a^2 + 2b^2 + 10ax + x^2 = 0$ Div:-	$4A - a^2 + 2b^2 =$ $10ax + x^2$	
36	$-4A - 2b^2 + 12a^2 + 12ax + 5x^2 = 0$ Div:-	$4A - 12a^2 + 2b^2 =$ $12a^2 + 5x^2$	
37	$-4A - 2b^2 + 5a^2 + 2ax + 5x^2 = 0$ Div:-	$4A + 2b^2 - 5a^2 =$ $2ax + 5x^2$	
38	$ec - e^2 - A + cx + dx - 2ex = 0$ Div : -	$A - ec + e^2 = x(c + d + 2e)$	Co.SK: $A - ec - e^2 = (a - e).(b - e) + dx + x^2$
39	$4ab - A + 2ax + 2bx + 0.25x^2 = 0$ Div:-	$A - (2a.2b) =$ $x2(a + b) + 0.25x^2$	
40.1	$9b^2 - 18A - 36a^2 - 36ax - 2.5x^2 = 0$ $(x = \text{diameter})$	$b^2 - 2A - 4a^2 =$ $4ax + 2.5x^2$	
40.2	$9b^2 - 18A - 36a^2 - 18ax - 2.5x^2 = 0$ $(x = 3 \text{ diameters})$		

41	$b^2 - 2A - 4a^2 +$ $4ax - 2.5x^2 = 0$ Div:-	$4a^2 + 2A - b^2 =$ $4ax - 2.5x^2$	
42	$4a^2 - c^2 - 2A +$ $4ax - 0.5x^2 = 0$ Div:-	$2A + c^2 - 4a^2 =$ $4ax - 0.5x^2$	
43	$[17500 \times 11a^2 +$ $1225 \qquad \times$ $628a^2 - 245000A] +$ $[17500 \times 22ax +$ $1225 \times 628ax] +$ $[17500 \times 11x^2 +$ $1225 \times 175x^2 +$ $61250 \times 3x^2] = 0$ Div:-	$1400A - 1099 \times$ $2a^2 - 1100a^2 =$ $1099 \times 4ax + 1100 \times$ $2a + 3249x^2$	
44	$25.6 - 200x = 0$ (trapezium)		$2 : c - c' \times d = C - C'$
45	$2a^2 - A - 2x^2 = 0$	$2x^2 = 2a^2 - A$ (equation reconstituted)	$a^2 - \frac{A}{2} = x^2$
46	$1.96A - (a^2 - 2x +$ $2.47x^2) = 0$ Div:-	$a^2 - 1.96A = 2ax -$ $2.47x^2$ xu yu	$1.96A - a^2 =$ $-2ax + 2.47x^2$
47	$A - (4ab + 2.8(a +$ $b)x + 0.96x^2) = 0$	$A - 4ab = 2.8(a +$ $b)x + 0.96x^2$	
48	$4a^2 - A - 4ax - bx = 0$	$4a^2 - A = 4ax - bx$	
49	$4a^2 - A +$ $5.6ax + 0.96x^2 = 0$ Div:-	$A - 4a^2 = 5.6ax +$ $0.96x^2$	
50	$4a^2 - 1.96A + 4ax -$ $0.96x^2 = 0$ Div:-	$1.96A - 4a^2 =$ $4ax - 0.96x^2$	
51	$A - 4a^2 +$ $5.6ax - 0.96x^2 = 0$ Div:-	$4a^2 - A = 4a \times$ $1.4x - 0.96x^2$	$\frac{2a^2 - A}{2} = 2 \times$ $1.4ax - 0.48x^2$

52	$4a^2 - 1.96A - 4ax - 0.96x^2 = 0$	$4a^2 - 1.96A = 4ax + 0.96x^2$	
53	$4a^2 - 1.96A - 4ax - 1.96x(2b - 2a) - 0.96x^2 = 0$	$4a^2 - 1.96A = 4ax - 1.96x(b-a) + 0.96x^2$	
54	$4a^2 - 1.96A + 4ax - 19.6x - 0.96x^2 = 0$ Div:-	$1.96A - 4a^2 = 4ax - 1.96(a-b)x - 0.96x^2$	
55	$ax - x^2 - 2A = 0$ Div:-	$2A = ax - x^2$	
56	$-14A + 44ax = 0$ Div:-	$14A = 44ax$	[1]: $\frac{22ax}{7} = A$ [2]: $[11(a^2 - a^2)] - 14A = 44ax$
57	$12a^2 - 4A + 12ax - 8x(a-b) - x^2 = 0$ Div:-	$4A - 12a^2 = 12ax - 4x(a-b) - x^2$	$4A - 3(2a)^2 = x6(a+b) - (a-b) - x^2$
58	$12a^2 - 4A - 12ax + 8x(a-b) - x^2 = 0$	$12a^2 - 4A = 12ax - 4x(a-b) + x^2$	
59	$A + S - [a^2 - \frac{3d^2}{4} + x^2] = 0$		$\frac{A+S}{19.25} = x^2$
60	$A + C - [\frac{3d^2}{4} - 9x^2 + \frac{3x^2}{4}] = 0$		$\frac{A+C}{10.5} = x^2$
61	$a^2 - 1.96A + 2.4ax - 0.03x^2 = 0$ Div:-	$1.96A - a^2 = 2 \times 1.2ax - 0.03x^2$	
62	$a^2 - 1.96A + 2.96ax + 0.2304x^2 = 0$ Div:-	$1.96A - a^2 = 2 \times 1.48ax + 0.2304x^2$	$\frac{49A}{25} - a^2 = 2 \times 1.48ax + 0.2304x^2$
63	$4A - [4(2a+i)^2 + 3(2a+2i)^2 + 16a^2] - x[16a + 8(2a+i) + 6(2a+2i)] - 8x^2 = 0$	$4A - 4(2a+i)^2 - 3(2a+2i)^2 - 16a^2 = x[16a + 8(2a+i) + 6(2a+2i)] - 8x^2$	

64	$1.96A - [2((a + d)^2 - 1.47a^2)] - x[4(a + d) + 4 \times 1.47a] - x^2 = 0$	[1]: $1.96A - 4(a + d)^2 + 1.47(2a)^2 = x[4(a + d) - 3 \times 1.96a] + x^2$ [2]: $1.96A - 4(a + d)^2 + 4 \times 1.47a^2 = 4x(a + d) - 4x \times 1.47a + x^2$	

Table A.2: The Distance Given in the Statement of the Problem

Chapter 1	Problem1	(Side − diameter)/2
	Problem2	(Side + diameter)/2
	Problem3	(Diagonal − diameter)/2
	Problem4	(Diagonal + diameter)/2
	Problem5	Perimeter − circumference
	Problem6	Side = circumference
	Problem7	Side − circumference
	Problem8	Perimeter + circumference
	Problem9	Perimeter + circumference + (side − diameter)/2
	Problem10	Perimeter + circumference + (diagonal − diameter)/2
	Problem11a	(Diameter − side)/2
	Problem11b	Diameter − diagonal
	Problem12	(Diameter + side)/2
	Problem13	(Diameter − diagonal)/2
	Problem14	(Diameter + diagonal)/2
	Problem15	Circumference − perimeter
	Problem16	Circumference = perimeter
	Problem17	Perimeter − diameter
	Problem18	Circumference + perimeter
	Problem19	Circumference + perimeter + (diameter − side)/2
	Problem20	Circumference + perimeter + (diameter − diagonal)/2

	Problem21	Side of the big square − side of the middle square = side of the middle square − side of small square
	Problem22	Diagonal − bisectrix of the triangular pond
Chapter 2	Problem23	Side − diameter
	Problem24	Side + diameter
	Problem25	Perimeter − circumference
	Problem26	Perimeter + circumference
	Problem27	Side − diameter
	Problem28	Perimeter − circumference
	Problem29	Perimeter + circumference
	Problem30	Diameter A + diameter B
	Problem31	(Diagonal + diameter)/2
	Problem32	diameter = length + width. Length − width
	Problem33	diameter − (length + width). Length − width
	Problem34	(Diameter − diagonal)/2. Length − width
	Problem35	(Diameter + diagonal)/2. Length − width
	Problem36	(Diameter − diagonal)2. Length + width
	Problem37	(Diameter + diagonal)/2. Length + width
	Problem38	Length A + width A. Length B + width B. Width A − width B.
	Problem39	(Length − diameter)2. (Width − diameter)/2.
	Problem40	(Diagonal − diameter)/2. Length + width. Length − width.
	Problem41	(Diagonal + diameter)/2. Length + width. Length − width.

	Problem42	(Diagonal + diameter)/2. Length − width.
Chapter 3	Problem43	Diameter of the middle circle = Diameter of the small circle + 9. Diameter of the big circle = diameter of the middle circle + 9.
	Problem44	2 different segments of the same length.
	Problem45	Distance from one corner of the outer square to the opposite corner of the inner square
	Problem46	Diagonal + diameter
	Problem47	(Length − diagonal)/2. (Width − diagonal)/2
	Problem48	(Side − length)/2. Length − width
	Problem49	(Side − diagonal)/2
	Problem50	(Diagonal − side)/2
	Problem51	(Side + diagonal)/2
	Problem52	(Diagonal + side)/2
	Problem53	(Diagonal + length)/2. (Diagonal + width)/2
	Problem54	(Diagonal − length)/2 (Diagonal − width)/2
	Problem55	Circumference A + circumference B + (Diameter A/2 − diameter B)
	Problem56	Diameter A/2 + diameter B
	Problem57	(Diameter− length)/2 (Diameter− width)/2
	Problem58	(Diameter + length)/2. (Diameter + width)/2
	Problem59	Side Diameter
	Problem60	Diameter Side
	Problem61	Diagonal − diameter − segment of diagonal
	Problem62	Diagonal − side − segment of diagonal

	Problem63	In small square: (Side − diameter)/2 Side of the small square + 50 = side of the big square Side of the big square + 50 = diameter
	Problem64	(Diagonal − diameter)/2. Circumference A− circumference B

Table A.3: Order of Transformation from Problem 1 to Problem 64

Shape:	Transformation:
a: square field, circular pond	A: removing corner(s) (*jian* 減).
b: circular field, square pond	B: stacking and unstacking areas (*tie* 貼, *die* 疊)
c: square field and square field	C: compensating areas (*bu* 補)
d: circular field, rectangular pond	D: expanding areas (*zhan* 展)
e: rectangular field, circular pond	E: multiplying by parts (*fen* 分)
f: square field, square pond	F: moving an area, cancelling areas (*qu⋯lai* 去⋯來, *luo* 漏)
g: square field, rectangular pond	
h: circular field, circular pond	

Roll 1	Problem	shape	transformation
	1	a	A
	2		B C
	3		A D
	4		B C D
	5		B E
	6a, 6b		A E
	7		A C E
	8		B E
	9		B E
	10		B E

	11a	b	B
	11b		B D
	12		B C
	13		A D
	14		B C D
	15		B E
	16		A E
	17		A C E
	18		B E
	19		B E
	20		B E
	21	three squares	B F
	22	Triangle in square	B D F
Roll 2	23	c	A E
	24		BB1
	25		AE
	26		BB2
	27		A E
	28		A E
	29		BB3
	30	two circles	BB4
	31	e	B C FF
	32	d	AA1
	33		B AA1
	34		B AA2
	35		AA2
	36		AA3
	37		A E F
	38	two rectangles	A E F
	39	e	A
	40		AA3 FF
	41		AA3 FF C
	42		AFC
Roll 3	43	three circles	A E
	44	trapezium	?
	45	f	AA3
	46	c	BB1 D
	47	Square in rectangle	A D1

48	g	B F
49	f	A D1
50		A D2
51		B C D1
52		B D2 F
53	g	BB2 D
54		A D2 FF
55	h	A
56		A E
57	d	A E
58		B E C
59	a and b	A
60	b and a	A E
61	a (in corner)	A D3
62	f (in corner)	A D3
63	a and c	A
64	a and h	A B C D

A.2   Classic of Computation of Master Sun, Division.

The *Classic of Computation of Master Sun* (194) explains how to divide 100 by 6 that is $100 \div 6 = 16\frac{4}{6}$. Here, the explanation appears in Hindu–Arabic numerals following Lam Lay-yong description [Lam and Ang (2004), 64]. The numerals are initially displayed as in [i]. The divisor 6 is then shifted two places to the extreme left [ii]. Then, since the division of 1 by 6 is not possible, the divisor 6 is shifted to the right [iii]. Because 6 is less than the '10' formed by the leftmost two digits of 100, the first digit of the quotient is in the tens place and should be placed above the tens of the dividend. One six is 6, so 1 is the quotient and 100 is reduced to 40 [iv]. Then, the divisor 6 is shifted to the right by one place [v]. Six sixes are 36, the quotient 6 is in the units place and 40 is reduced to 4 [vi]. The remainder 4 is called *zi* 子, 'numerator' (literally 'son') and the divisor 6 is called *mu* 母, 'denominator' (literally 'mother').

```
shang ┌─────┐   ┌─────┐   ┌─────┐   ┌─────┐   ┌─────┐   ┌─────┐ shang
 shi  │1 0 0│ → │1 0 0│ → │1 0 0│ → │  1  │ → │  1  │ → │ 1 6 │  zi
 fa   │    6│   │6    │   │    6│   │ 4 0 │   │ 4 0 │   │   4 │  mu
      └─────┘   └─────┘   └─────┘   │    6│   │    6│   │    6│
        [i]       [ii]      [iii]   └─────┘   └─────┘   └─────┘
                                      [iv]      [v]       [vi]
```

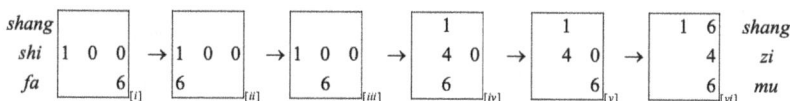

A.3 Classic of Computation of Master Sun, the Extraction of a Square Root.

In Chapter 2 (中卷), problem 19 of the *Classic of Computation of Master Sun* states[29]:

今有積，二十三萬四千五百六十七步。

問：為方幾何？

答曰：四百八十四步九百六十八分步之三百一十一。

Suppose that there is an area of two hundred thirty-four thousand five hundred sixty-seven bu^2. (234, 567 bu^2)

The problem is as follows: how long is the side of the square?

Answer: four hundred eighty-four *bu*, and three hundred eleven parts of nine hundred sixty-eight *bu*. ($484\frac{311}{968}$ *bu*)

術曰: 置積二十三萬四千五百六十七步，為實，

The procedure says the following: One places the area two hundred thirty-four thousand, five hundred sixty-seven *bu* as a dividend (*shi*).

shi	2	3	4	5	6	7	(step 1)

次借一算為下法，步之超一位至百而止。

Next, one borrows one rod as the lower divisor (*xia fa*). [From the place of the] *bu* (i.e. units), one passes over one place (*chao yi wei*) to reach the hundred[30] and stops.

---

[29] Text from: http://ctext.org/sunzi-suan-jing/zh. The editors' commentaries have not been reproduced. The literal translation is original. The representation of the counting support is a hypothetical reconstruction by Lam Lay-Yong in [Lam and Ang (2004)] but the actual form of these materials remains speculative.

[30] In this last sentence, the word 'hundred' was replaced by 'ten thousand' by Lam Lay-Yong. [Lam and Ang (2004), 95] justified this correction with the reading of steps 7 and 12. The reader is asked to move this rod backwards to the right two places at a time so that ultimately it is once again below the unit of the shi. In the explanation of the algorithm for root extraction that in the *Nine Chapters*, Chemla shows that the borrowed rod (*jie fa*) is first placed in the position of the units. This rod is then moved towards the left, from 10^2 to 10^2 until it reaches the farthest position under the dividend, i.e. 10^{2n} if the first number of the root is 10^n. From these n 'jumps', one deduces the first number of the root, called the quotient [Chemla and Guo (2004), 326]. Although the algorithm of

shi	2	3	4	5	6	7
xiafa						1

shi	2	3	4	5	6	7
xiafa		1				

(step 2)

上商置四百于實之上，

One places four hundred [as] the quotient (*shang*) above the dividend.

shang			4			
shi	2	3	4	5	6	7
xiafa		1				

(step 3)[31]

副置四萬于實之下，下法之上，名為方法；

Next, one places forty thousand below the dividend (*shi*) and above the lower divisor (*xia fa*) and one calls it the square divisor (*fang fa*).

shang			4			
shi	2	3	4	5	6	7
fangfa		4				
xiafa		1				

(step 4)[32]

命上商四百除實，

One names (*ming*[33]) four hundred as the quotient (*shang*) in the upper [position] and removes (*chu*) from the dividend (*shi*).

---

the *Nine Chapters* is slightly different from that of the *Computational Classic of Master Sun*, the interpretation of the role of the 'borrowed rod' in the *Computational Classic of Master Sun* and its correction requires more discussion. [Chemla (1994c), 17] in a comparison of the algorithm of root extraction in the Computational *Classic of Zhang Qiujian* with the one by Kushyar ibn Labban, explains the role of the 'borrowed rod'. This rod has different roles in Chinese algorithms for root extraction.

[31] According to Lam Lay-Yong, the determination of the fourth digit for the hundreds of the root occurs through trial and error to find the largest possible digit that leaves a non-negative numeral for the dividend. Lam Lay-Yong formulates the same hypothesis for steps 8 and 13, but this cannot be confirmed.

[32] Having obtained the digit for the hundreds of the root, the same digit is also placed in the row immediately below the *shi* in the same column as that of the single rod of the *xia fa*. It represents 40,000 for the *fang fa*.

[33] In the lexicon to the *Nine Chapters*, Chemla translates *ming* as 'to name'. This term applies when the operation for the root extraction is not finished after one has just determined the number of the unit of the root; the result is produced by 'naming' the number for which the root is sought. [Chemla and Guo (2004), 963]; [Li (1990), 150].

shang			4			
shi	7	4	5	6	7	
fangfa	4					
xiafa	1					

(step 5)[34]

除訖，倍方法，

[Once] the removal is completed, one doubles the square divisor (*fang fa*).

shang			4			
shi	7	4	5	6	7	
fangfa	8					
xiafa	1					

(Step 6)

方法一退，下法再退，

One shifts (*tui*) the square divisor (*fang fa*) [to the right] by one [place] and shifts the lower divisor (*xia fa*) again.

shang			4			
shi	7	4	5	6	7	
fangfa		8				
xiafa			1			

(step 7)[35]

復置上商八十以次前商，

Next, one places the quotient (*shang*) in the upper [position], 80, next to the previous quotient (*shang*).

shang			4	8		
shi	7	4	5	6	7	
fangfa		8				
xiafa			1			

(step 8)

副置八百于方法之下，下法之上，名為廉法；

One also places eight hundred below the square divisor (*fang fa*) and above the lower divisor (*xia fa*) and calls it the edge divisor (*lian fa*).

---

[34]Sunzi uses the quotient to multiply the square divisor. The place value of the product corresponds with the digit of the square divisor so that the product, 16, is subtracted from the dividend, 23. The subtraction of these two numbers leaves 7 in place of 23, so the dividend is now 74, 567.

[35]The single rod of the low divisor (*xia fa*) in the hundreds' place indicates the determination of the tens of the root.

shang			4	8	
shi	7	4	5	6	7
fangfa		8			
lianfa		8			
xiafa		1			

(step 9)[36]

方廉各命上商八十以除實，

One [multiplies] each of the square (*fang*) and edge (*lian*) [divisors together], names (*ming*) 80 as the quotient (*shang*) in the upper [position] and removes this from the dividend (*shi*).

shang			4	8	
shi	1	0	5	6	7
fangfa		8			
lianfa		8			
xiafa		1			

shang			4	8	
shi		4	1	6	7
fangfa		8			
lianfa		8			
xiafa		1			

(step 10)[37]

除訖，倍廉法，上從方法，

[Once] the removal is complete, one doubles the side divisor (*lian fa*) and joins the square divisor (*fang fa*) above.

shang			4	8	
shi		4	1	6	7
fangfa		9	6		
xiafa		1			

(step 11)

方法一退，下法再退，

---

[36]Having obtained the number for the tens of the root, the same number is placed in the row immediately below the *fang fa* and in the same column as that of the single rod of the *xia fa*.

[37]The tens digit of the *shang* multiplies the value in the *fang fa*. The product, 64, is subtracted from the 74 of the *shi*. The subtraction of the two numbers leaves 10 in place of 74, so the *shi* is now 10,567. Next, the tens of the *shang* multiplies the digit of the *lian fa*. The product, 64, is subtracted from the 105 of the *shi*. The subtraction of the two numbers leaves 41 in place of 105, so the *shi* is now 4,167.

One shifts the square divisor (*fang fa*) [to the right] by one [place] and shifts the lower divisor (*xia fa*) again.

shang			4	8	
shi	4	1	6	7	
fangfa			9	6	
xiafa				1	

(step 12)[38]

復置上商四以次前，

Next, one places the quotient (*shang*) in the upper [position], 4, next to the previous one.

shang			4	8	4
shi	4	1	6	7	
fangfa			9	6	
xiafa				1	

(step 13)

副置四于方法之下，下法之上，名曰隅法；

One also places 4 below the square divisor (*fang fa*) and above the lower divisor (*xia fa*) and calls it the corner divisor (*yu fa*).

shang			4	8	4
shi	4	1	6	7	
fangfa			9	6	
yufa				4	
xiafa				1	

(step 14)

方廉隅各命上商四以除實，

One [multiplies] each square (*fang*), side (*lian*) and corner (*yu*); one names four as the quotient (*shang*) in the upper [position] and removes it from the dividend (*shi*).

shang		4	8	4
shi		5	6	7
fangfa		9	6	
yufa				4
xiafa				1

---

[38] The single rod of the *xia fa* in the units' place indicates the determination of the units' digit for the root.

shang	4	8	4
shi	3	2	7
fangfa	9	6	
yufa			4
xiafa			1

shang	4	8	4
shi	3	1	1
fangfa	9	6	
yufa			4
xiafa			1

(step 15)[39]

除訖，倍隅法，從方法，

[Once] the removal is complete, one doubles the corner divisor (*yu fa*) and joins this to the [other] corner divisor (*fang fa*).

shang	4	8	4
shi	3	1	1
fangfa	9	6	8
xiafa			1

(step 16)[40]

上商得四百八十四，下法得九百六十八，不盡三百一十一，是為方四百八十四步九百六十八分步之三百一十一。

The quotient (*shang*) in the upper [position] is four hundred eighty-four, the divisor (*fa*) in the lower [position] yields nine hundred sixty-eight and the remainder (*bu jin*, lit. 'not exhausted') is three hundred eleven. This makes the side of the square four hundred eighty-four *bu*, three hundred eleven parts out of nine hundred sixty-eight *bu*. ($484\frac{311}{968}$ *bu*).

---

[39]The digit for the units of the *shang* multiplies the digits of the *fang fa*, including *lian fa*. Following the method of multiplication, 4 first multiplies 9 to give 36 and this is subtracted from 41 above to leave 567 for the *shi*. Next, 4 multiplies 6 to give 24 and this is subtracted from the 56 above to leave 327 for the *shi*. The digit for the units of the *shang* also multiplies the digit of the *yu fa*. The product, 16, is subtracted from the 27 of the *shi*. The subtraction of the two numbers leaves 11 in place of 27, so the *shi* is now 311.

[40]This step is like steps 6 and 11.

A.4 Web of Problems

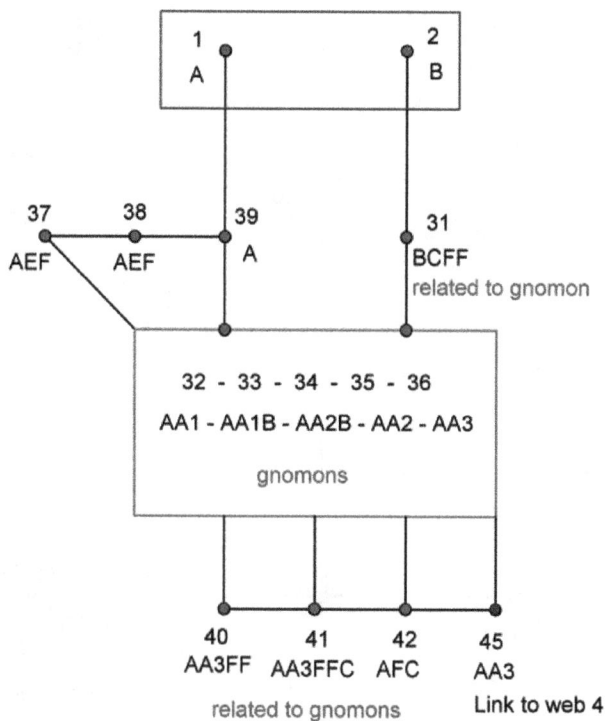

Fig. S.1 Graph 3 of Problems 31–45

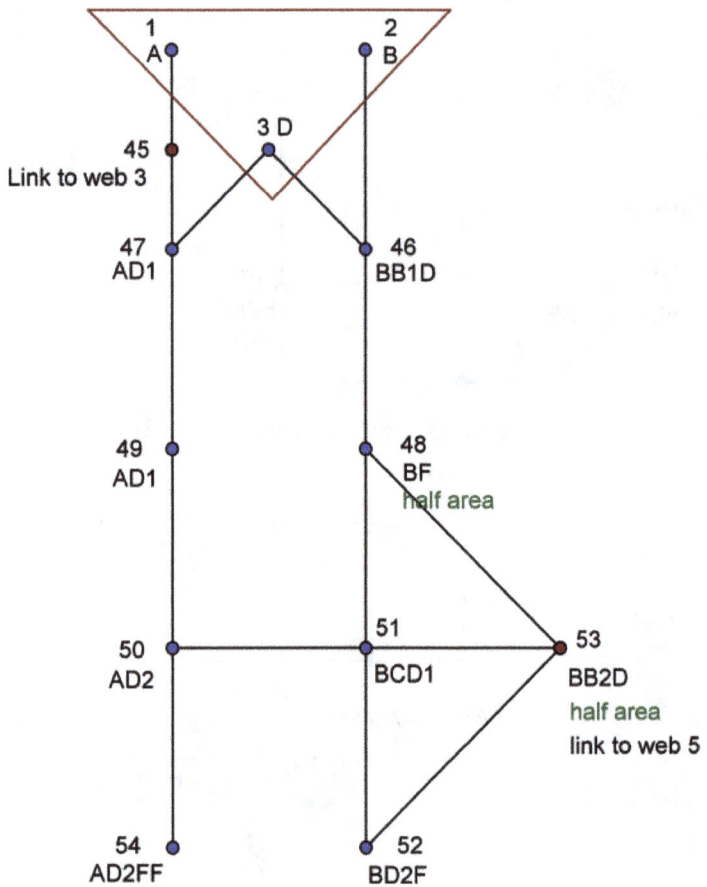

Fig. S.2   Graph 4 of Problems 45–54

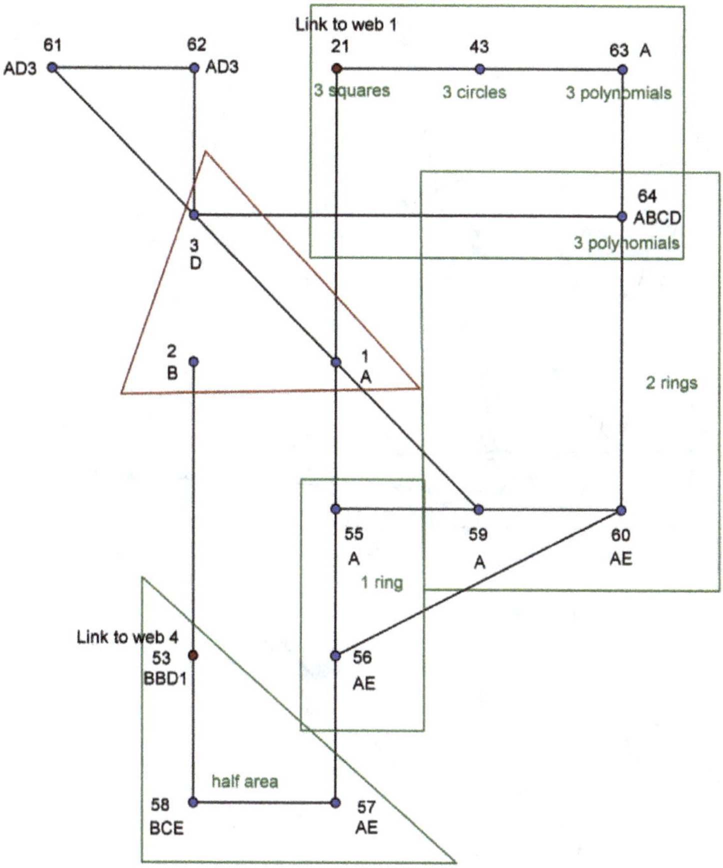

Fig. S.3 Graph 5 of Problems. 55–64

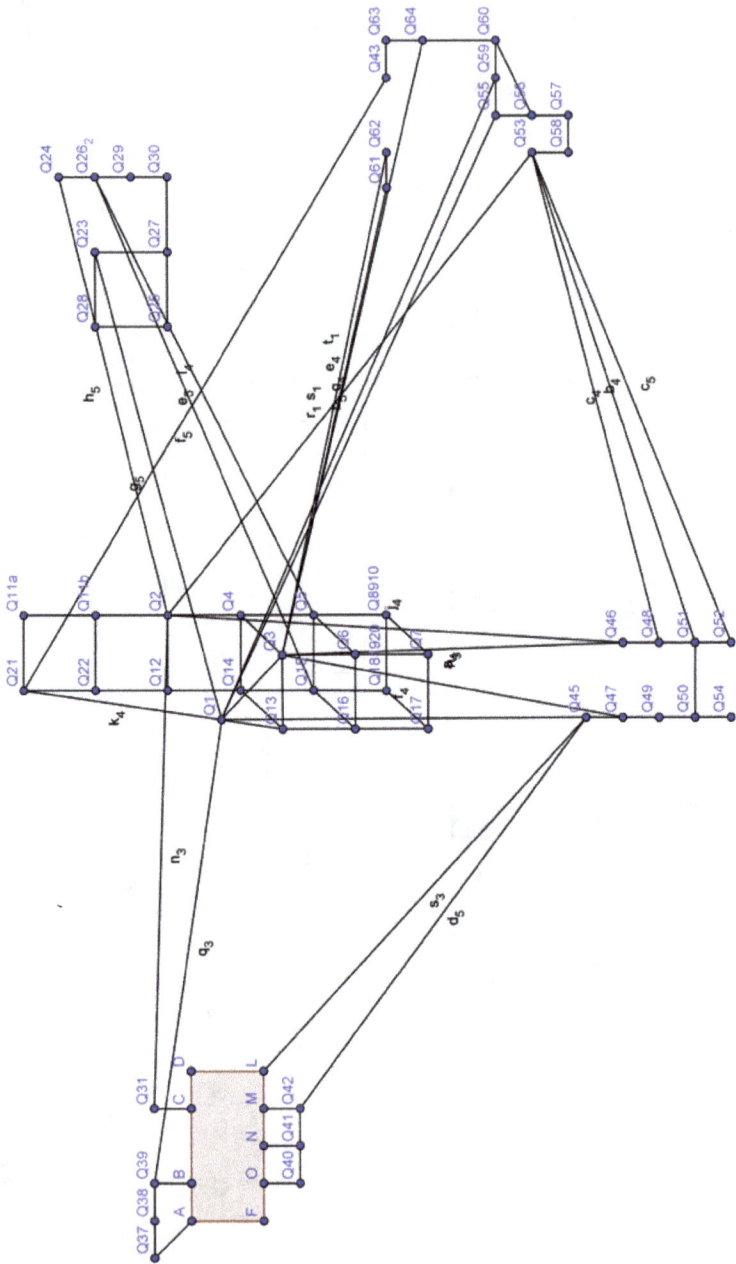

Fig. S.4   General Graph of Dependencies among Problems

Appendix B

Chinese Glossary

- An 案 commentary
- Bianjing 汴京
- Bishu Congxue 壁書叢削 *Amendments of Books on the Wall Shelves*
- Bu 補 compensating' or 'mending' areas
- Cao 草
- Ceyuan Haijing 測圓海鏡 *Sea Mirror of Circle Measurement*
- Chang Fa 常法 constant divisor
- Chouren Zhuan 疇人傳 *The Inventory of Biographies of Scientists*
- Cong 從 adjunct
- Da Yan 大衍 lit, 'great extention'
- Da 答 answer
- Dai Cong Kai Fang 帶從開方
- Daode Jing 道德經
- De 得 to yield
- Die 疊 to stack, to pile up
- Duan 段 pieces [of area]
- Fa 法 divisor
- Fan Shuo 泛說 supernumerary talks
- Fang Fa 方法
- Fenglong Mountains 封龍山
- Fen 分 multiplying by parts
- Fu Cong 負從 negative adjunct
- Fuyu 負隅 negative corner
- Fu 負 negative
- Gen 根 the root
- Gou 勾 base
- Gu 股 height

235

- He Bu 和步 sum
- Hebei 河北
- Jian Cong 減從 subtract the adjunct
- Jian Ji 見積 real area
- Jian 減 subtract
- Jian 減 removing corner(s)
- Jie Gen Fang 借根方 borrowing the root and powers
- Jin You 今有 let us suppose
- Jingzhai Wenji 敬齋文集 *Collection of Works by Jingzhai*
- Jingzhai Gujing Tu 敬齋古今黈 *Commentary of Jingzhai on Things Old and New*
- Jiu Shu 舊術 old procedure
- jiu Zhang Suanshu 九章算術 *Nine Chapters on Mathematical Procedures*
- Ji 積 area as product
- Kai Ping Fang Chu 開平方除 the extraction of the square root by division
- Kai Ping Fang 開平方 opening the square
- King Wen 文王
- Kong 空 void
- Li Fang 立方 the cube
- Lian 廉 edge
- Linan 臨安
- Mi 冪 surface
- Mu 母 denominator
- Nei Dan 內丹 inner alchemy
- Ping Fang 平方 square
- Ping Yang 平陽
- Qu 去 to go
- Quan Zhen Jiao 全真教 School of Complete Perfection
- Luo 漏 moving an area, cancelling areas
- Shang 商 quotient
- Shanxi 山西
- Shi 實 dividend
- Shi 式 configuration
- Shu 術 procedure
- Shushu Jiuzhang 數書九章 *Mathematical Treatise in Nine Chapters*
- Si Yuan Shu 四元術 *Procedure of Four Sources*

- Si Yuan Yu Jian 四元玉鑑 *The Precious Mirror of Four Sources*
- Siku Quanshu 四庫全書 *Complete Library of the Four Treasuries*
- Sunzi Suanjing 孫子算經 *Classic of Computation of Master Sun*
- Tai 太
- Tian Yuan 天元 celestial source
- Tianmu Bilei Chengchu Jiefa 田畝比類乘除捷法 *Fast Methods of Multiplication and Division Related to [various] Categories of Fields and [their] Measures*
- Tiao Duan 條段 section of pieces [of areas]
- Tie 貼 to paste
- Tongwen Guan 同文舘 Beijing School of Foreign Language
- Tu 圖 diagram
- Wei 為 as, to make
- Wen 問 question
- Xiang Xiao 相消 eliminated from one another
- Xiangjie Jiu Zhang Suanfa 詳解九章算法 *Detailed Analysis of the Mathematical Methods in the 'Nine Chapters'*
- Xu 虛 empty
- Yang 陽
- Yang Hui Suan Fa 楊輝算法 *Yang Hui's Computing Methods*
- Yi Yu 益隅 augmented corner
- Yi 益 augment
- Yi 義 meaning
- Yigu Genyuan 議古根源 *Discussion on the Origin of Ancient Methods*
- Yigu Suanfa 益古算法 *Computing Method Augmenting the Ancient [knowledge]*
- Yigu Yanduan 益古演段 *Development of Pieces [of Areas] [according To] [the Collection] Augmenting the Ancient [knowledge]* (shortened as *Development of Pieces [of Areas]*)
- Yijing 易經 *Book of Changes*
- Yin 陰
- Yingyong Suanfa 應用算法 *The Computing Method for Application*
- Yongle Dadian 永樂大典 *Great Canon of Yongle*
- Yuan Shi 元史 official history of the Yuan
- Yuan 元
- Zhan 展 expanding areas
- Zhejiang 浙江
- Zhen Shu 真數 is real quanity (constant)
- Zhengyu 正隅 positive corner

- Zheng 正 positive
- Zhibuzu Zhai Congshu 知不足齋叢書 *Collected Works of the Private Library of Knowing Our Own Isufficiencies*
- Zhizhai Shu Lu Jieti 直齋書錄解題 *Explanations for the Titles in the Zhizhai Book List*
- Zhongshan 中山
- Zi 子 numerator

Appendix C

Chinese Names

- Chen Zhensun 陳振孫
- Cheng Dawei 程大位
- Dai Zhen 戴震
- Fu Xi 伏羲
- Jiang Shunyuan 蔣舜元
- Jiang Zhou 蔣周
- Jiaqing 嘉慶
- Kangxi 康熙
- Kubilai Khan 忽必烈
- Li Chunfeng 李淳風
- Li Rui 李銳
- Li Shanlan 李善蘭
- Li Ye 李冶 literary name Renqing 仁卿, appellation Jingzhai 敬齋
- Liu Hui 劉徽
- Mei Juecheng 梅瑴成
- Qianlong 乾隆
- Qin Jiushao 秦九韶
- Ruan Yuan 阮元
- Shang Zhou Wang 商紂王
- Shao Yong 邵雍
- Tang Gaozong 唐高宗
- Wang Lai 汪萊
- Yang Hui 楊輝
- Zhou Wen Wang 周文王
- Zhu Shijie 朱世杰
- Zu Yi 祖頤

Bibliography

Primary Sources:

Li Ye 李冶 [1248]. 测圆海镜 (Ceyuan haijing [*Sea Mirror of Circle Measurements*])

Li Ye 李冶 [1259]. 益古演段 (Yigu yanduan [*The Development of Pieces of Areas according to the Collection Augmenting the Ancient Knowledge*])

Both edited in:

(1) Edition of 1789, 文渊阁四库全书, (Wen yuan ge Siku quanshu [*Complete Library of the Four Treasuries, Wenyuan Pavilion*]), original edition from National Palace Museum, Taiwan.

(2) Edition of 1789, 文津阁四库全书, (Wen jin ge Siku quanshu [*Complete Library of the Four Treasuries, Wenjin Pavilion*]), Vol. 799. Reprint. 2005.

(3) Edition of 1789, 知不足斋丛书 (Zhibuzu zhai congshu [*Collected Works of the Private Library of Knowing Our Own Insufficiencies*]), reprint in 中国科学技术典籍通汇: 数学篇 (Zhongguo kexue jishu dianji tong hui: Shuxue pian [Source Materials of Ancient Chinese Science and Technology: Mathematics Section]). 郭书春, Guo Shuchun (ed). 河南教育出版社 (Henan jiayu chubanshe [Henan Education Press]). 1993. Vol. 1.

Yang Hui 杨辉 [1275]. 杨辉算法, Yang Hui suanfa [*Yang Hui Computational Methods*]. 2nd book: 田亩比类乘除捷法, Tian mu bilei cheng chu jie fa [*Fast Methods of Multiplication and Division Related to Various Categories of Fields and Their Measures*]. 1433 Manuscript preserved in Beijing library based on a 1378 Ming edition, reprint in Guo Shuchun (ed) 郭书春中国科学技术典籍通汇: 数学卷 (Zhongguo kexue jishu dianji tong hui: Shuxue juan [Chinese Science and Technology Books Collectanea: Mathematics Volume]). 河南教育出版社 (Henan jiaoyu chubanshe [Hunan educational press]) 1993.

Zhu Shijie 朱世杰 [1303]. 四元玉鉴, si yuan yu jian [*The Jade Mirror of the Four Unknowns*].Translated into English by Chen Tsai Hsin (Chen Zaixin) 陈在新, 1925, reedited and completed by Guo Shuchun 郭书春, and Guo Jinhai 郭金海. 大中华文库, 辽宁教育出版社 (da zhonghua wenku [*Library of the Chinese Classics*], Liaoning jiaoyu chubanshe [Liaoning Educational Press]). 2006.

Qin Jiushao 秦九韶 [1247]. 数学九章, shuxue jiu zhang [*Mathematical Treatise in Nine Chapters*]. 文津阁四库全书 (Wen jin ge Siku quanshu [*Complete Library of the Four Treasuries, Wenjin Pavilion* **798**]). Reprint. 2005.

Ruan Yuan 阮元 [1799]. 畴人传 (Chou ren zhuan [*Biographies of Mathematicians*]), Vol. 82. 中华汉语工具书书库 (Zhonghua hanyu gongju shu shuku [*Collection of Chinese Reference Works*]). 安徽教育出版社 (Anhui jiaoyu chubanshe [Anhui Education Press]). 元史, Yuan shi [*Official History of the Yuan*], 1370. 四库全书, 文津阁 Wenjinge siku quanshu. Reprint. 2005. Vol. 290. Ch. 160, 11–13.

Secondary Sources:

Ang (Tian Se) [1978]. Chinese interest in right angle triangles, *Historia Mathematica* **5** 253–266.

Benoit (Paul), Chemla (Karine) and Ritter (Jim) [1992]. Histoire de fraction, fraction d'histoire. Birkhauser Verlag.

Bray (Francesca) [2007]. Introduction: the powers of Tu. Bray. F, Dorofeeva-Lichtmann. V, Metailie. G (eds.) Graphics and Text in the Production of Technical Knowledge in China. Sinica Leidenisa **79**. Brill.

Bréard (Andrea) [2000]. La recomposition des mathématiques chez Zhu Shijie: la constitution d'un domaine spécifique autour du nombre 'quatre'. *Oriens Occidens* **3** 259–277.

Bréard (Andrea) [2012]. Divination with hexagrams as combinatorial practice. A paradigmatic model in mathematics. *Zhouyi Studies* (English Version) **8** 157–174.

Chemla (Karine) [1982]. Etude du livre 'reflets des mesures du cercle sur la mer' de Li Ye. Thèse de doctorat de l'université Paris XIII. Not published.

Chemla (Karine) [1993]. 李冶《测圆海镜》的结构及其对数学知识的表述 (Li Ye, Ceyuan haijing de jieguo yu qi dui shuxue zhishi de biaoshu [The structure of Li Ye's *Sea Mirror of Circle Measurements* and its descriptions twoards mathematical knowledge]), 《数学史研究文集第五辑》(shuxue shi yanjiu di wu ji [Research on History of Mathematics Series **5** 123–142.]).

Chemla (Karine) [1994a]. Different concepts of equations in *The Nine Chapters on Mathematical Procedures* 九章算术 and in the Commentary on it by Liu Hui (3rd century). *Historia Scientiarum* 4(2) (1994) 113–137.

Chemla (Karine) [1994b]. Essais sur la signification mathématique des marqueurs de couleur chez Liu Hui (3eme sciècle). A. Peyraube, I. Tamba, A. Luca (eds), *Mélanges en hommage à Alexis Rygaloff, Cahiers de linguisitique. Asie Orientale* **23**(1) 61–76. http://www.persee.fr

Chemla (Karine) [1994c]. Similarities between Chinese and Arabic mathematical writings: (I) root extraction. *Arabic Sciences and Philosophy* **4**(02) 207–266.

Chemla (Karine) [1994d]. Nombres, opérations et équations en divers fonctionnements: quelques méthodes de comparaison entre des procédures élaborées dans trois mondes différent. I. Ang and P. E Will (eds). Nombres, astres, plantes et viscères. Sept essais sur l'histoire des sciences et des techniques en Asie orientale 1–36. Paris. Collège de France, Institut des Hautes Etudes Chinoises (Mémoires de l'Institut des Hautes Etudes Chinoises XXXV).

Chemla (Karine) [1995]. Algebraic equations east and west until the middle ages. K. Hashimoto, C. Jami, L. Skar (eds.), *East Asian Science: Tradition and Beyond* 83–89: Papers from the Seventh International Conference on the History of Science in East Asia, Kyoto, 2–7 August 1993, Kansai University Press, Osaka.

Chemla (Karine) [1996]. Positions et changements en mathématiques à partir des textes chinois des dynasties Han à Song-Yuan. Quelques remarques. *Extrême-Orient, Extrême-Occident* **18** 115–147.

Chemla (Karine) [1997a]. Croisements entre réflexion sur le changement et pratique des mathématiques en Chine ancienne : le cas des Neuf chapitres sur les procédures mathématiques et leurs commentaires. J. Gernet et M. Kalinowski (ed), En suivant la Voie royale, Mélanges offerts en hommage à Léon Vandermeersch (Etudes thématiques) 191–205. Ecole Française d'Extrême Orient.

Chemla (Karine) [1997b]. Fractions and irrationals between algorithm and proof in ancient China. *Studies in History of Medicine and Science. New Series* **15** 31–54.

Chemla (Karine) [2000]. Les problèmes comme champs d'interprétation des Algorithmes dans les Neuf Chapitres sur les procédures mathématiques et leurs commentaires. De la résolution des systèmes d'équations linéaires. Oriens-Occidens. Sciences, mathématiques et philosophie de l'antiquité à l'âge classique **3** 189–234.

Chemla (Karine) [2001]. Variété des modes d'utilisation des Tu dans les textes mathématiques des Song et Yuan. Pre-print. Conference from image to action: the function of Tu-Representation in east asian intellectual culture, Paris, 2001. http://hal.ccsd.cnrs.fr, section Philosophie, sub-section 'histoire de la logique et des mathématiques'.

Chemla (Karine) and Guo (Shuchun) 郭书春 [2004]. Les neuf chapitres. Le classique mathématique de la Chine ancienne et ses commentaires. Paris. Dunod.

Chemla (Karine) [2004]. What is the content of this book? A plea for developing history of sciences and history of text conjointly. Chemla Karine (ed), History of Science, History of Text. Springer, Dordrecht.

Chemla (Karine) [2005]. Geometrical figures and generality in ancient China and beyond: Liu Hui and Zhao Shuang, Plato and Thabit ibn Qurra. *Science in Context* **18** 123–166.

Chemla (Karine) [2006a]. Artificial language in the mathematics of ancient China. *Journal of Indian Philosophy* **34** 31–56.

Chemla (Karine) [2006b]. La généralité, valeur épistémologique fondamentale des mathématiques de la Chine ancienne. Journée Jean Filliozat, *Comptes rendus de l'Académie*, Académie des Inscriptions et Belles-Lettres, Institut, 2006 (2008), 10 bis, fascicule IV, 113–146.

Chemla (Karine) [2010]. Changes and continuities in the use of diagrams Tu in Chinese mathematical writings (third century to fourtheen century) [1]. *East Asian Science, Technology and Society* **4**(2) 303–326.

Chemla (Karine) [2016]. The dangers and promises of comparative history of science. *Sartoniana* 174–198. Halshs-01164229.

Chia (Lucile) [2002]. Printing for Profit. The Commercial Publishers of Jianyang Fujian (11th–17th Centuries). Harvard-Yenching Institute Monograph Series 56.

Cullen (Christopher) [1996], Astronomy and mathematics in ancient China: the Zhou bu suan jing, Needham Research Institute Studies **1**. Cambridge University Press.

Dauben (Joseph) [2007]. Chinese Mathematics. Victor Katz (ed), The Mathematics of Egypt, Mesopotamia, China, India and Islam, A Source Book. Princeton University Press. Princeton and Oxford. 187–380.

Eco (Umberto) [2004]. Mouse or Rat: Translation As Negotiation. Orion Pub Co.

Eifring (Halvor) [2016]. Asian Tradition of Meditation. University of Hawaii Press.

Giaquinto (Marcus) [2008]. Visualizing in mathematics. Mancosu (Paolo) (ed), The Philosophy of Mathematical Practice. Oxford University Press.

Guo (Shuchun) 郭书春 [1982]. 《九章算术》中的整数勾股形研究 (jiu zhang suanshu zhong de zhengshu gou gu xing yanjiu [A study on the right-angled triangles with integral side lengths in the *Nine Chapters* on mathematical procedures]); 科技史文集, 第 8 辑 (keji shi wenji, di 8 ji) 54–66.

Guo (Shuchun) 郭书春 [1991]. 中国古代数学 (Zhonguo gudai shuxue [*Mathematics in Ancient China*]). 山东教育出版社 (Shandong jiaoyu chubanshe [Shandong Education Press]). 1–174.

Guo (Shuchun), 郭书春 [2010]. 中国科学技术史. 数学卷. (Zhongguo kexue jishu shi: Shuxue juan [History of Chinese Science and Technology: Mathematics Volume]). 科学出版社, 北京. (Kexue chubanshe [Beijing: Science Press]).

Guo (Xihan) 郭熙汉 [1996]. 杨辉算法导读 (Yang Hui suanfa daodu [*Introduction to Yang Hui suanfa*]). 中华传统数学名著导读丛书. 李迪 (ed). (zhonghua chuantong shuxue mingzhu daodu congshu [*Collection of masterpieces in Chinese traditional mathematics*]), 湖北教育出版社 (Hubei jiaoyu chubanshe, [Hubei educational press]).

Hart (Roger) [2011]. The Chinese Roots of Linear Algebra. The Johns Hopkins University Press. Baltimore.

Ho Peng Yoke [1973]. Li Chih. Charles Coulston Gillispie (ed), Dictionary of Scientific Biography. Scribner's Sons. **8** 313–320.

Hoe (Jock) [1977]. Les systèmes d'équations polynômes dans le Siyuan Yujian (1303). Paris, Collège de France, Mémoires de l'Institut des Hautes Etudes Chinoises **6**.

Hoe (Jock) [2008]. The Jade Mirror of the Four Unknowns by Zhu Shijie. Mingming bookroom. Christchurch. New Zealand.

Horiuchi (Annick) [2000]. La notion de yanduan: quelques réflexions sur les méthodes "algébriques" de résolution de problèmes en Chine aux Xe et XIe siécles, *Oriens–Occidens* **3** 235–258.

Horng (Wann-Sheng) 洪万生 [1993a]. Chinese mathematics at the turn of the 19th century : Jiao Xun, Wang Lai and Li Rui. Cheng-hung Lin and Daiwei Fu (Eds), *Philosophy and Conceptual History of Science in Taiwan*, 167–208.

Horng (Wann-Sheng) 洪万生 [1993b]. 談天三友焦循、汪萊與李鋭：清代經學與算學關係試論, 洪萬生（編）, 談天三友 (Tan tian san you [*Three Friends Talking about the Heavens*]), 37–124, 台北: 明文書局 (Taibei: Ming-Wen shuju [Taipei: Ming-Wen Bookstore]).

Horng (Wann-Sheng) 洪万生 [1999]. 全真教与金元数学 — 以李冶 (1192–1279) 为例. (Quanzhenjiao yu Jin Yuan shuxue — yi Li Ye (1192–1279) wei li [*The School of Complete Perfection and the Mathematics of the Jin and Yuan dynasties*]). Wang Qiugui 王秋桂 (ed.) 金庸小说国际学术研讨会论文集 (Jin Yong xiaoshuo guoji xueshu yantaohui lunwen ji [*Proceedings of the International Confrence on Jin Yong's Novels* 67–83]). 台北源流出版公司 (Tai bei: Yuanliu chuban gongsi).

Jami (Catherine) [2012]. The Emperor's New Mathematics: Western Learning and Imperial Authority during the Kangxi Reign (1662–1722). Oxford, Oxford University Press.

Katz (Victor J.) and Hunger Parshall (Karen) [2014]. Taming the Unknown. A History of Algebra from Antiquity to the Early Twentieth Century. Princeton University Press. Princeton and Oxford.

Knobloch (Eberhard) [2016]. Generality in Leibniz's Mathematics. The Oxford Handbook of Generality in Mathematics and the Sciences. Chemla. K, Chorlay. R, Rabouin. D (Eds.) Oxford University Press.

Knuth (Donald E.) [2013]. Two Thousand Years of Combinatorics. Wilson Robin and Watkins John J.(eds) Combinatorics: Ancient and Modern. Oxford Univesity Press.

Kong (Guoping) 孔国平 [1988]. 李冶传 (Li Ye zhuan [Biography of Li Ye]). 北京: 河北教育出版社 (Beijing: Hebei jiaoyu chubanshe [Beijing: Hebei education press]).

Kong (Guoping) 孔国平 [1996]. 测圆海镜导读 (Ceyuan haijing daodu [Introduction to Ceyuan Haijing]). 中华传统数学名著导读丛书 (Zhonghua chuantong shuxue mingzhu daodu congshu [*Collection of masterpieces in Chinese traditional mathematics*]), 湖北教育出版社 (Hubei jiaoyu chubanshe [Hubei Education Press]).

Kong (Guoping) 孔国平 [1999]. 李冶朱世杰与金元数学 (Li Ye Zhu Shijie yu jinyuan shuxue [*Li Ye, Zhu Shijie and Jin-Yuan Mathematics*]), 中国数学史大系 (Zhongguo shuxueshi daxi [*Collection of History of Chinese Mathematics*]), 河北科学技术出版社 (Hebei kejishu chubanshe [Hebei Science and Technology Press]).

Lam (Lay-Yong) [1977]. A Critical Study of Yang Hui Suanfa, a Thirteen Century Chinese Mathematical Text. Singapore University Press.

Lam (Lay-Yong) [1984]. Li Ye and his Yi Gu Yan Duan (old mathematics in Expanded Sections). *Archive for history of exact sciences* **29**(3) 237–266.

Lam (Lay-Yong) Ang (Tian Se) [2004]. Fleeting Footsteps, Tracing the Conception of Arithemtic and Algebra in Ancient China. World Scientific, Singapore.

Lan (Lihong) 蓝丽蓉 and Hong (Tianxi) 洪天锡（赐）[1985]. 孟宪福译：李冶和《益古演段》(Mengxianfa yi: Li ye he "yigu yan duan" [*Li Ye and the Yigu yanduan*]),《科学史译丛》1985 年第 4 期，第 11–12 页 (Kexueshi yicong [*History of Sciences Collection of Translation*] 1985(4) 11–12).

Li (Di) 李迪 [1997]. 天元术与李冶 (tian yuan shu yu Li Ye [*The Tianyuan shu and Li Ye*]). 中国数学通史. 宋元卷 (zhongguo shuxue tongshe. Song–Yuan juan [*Chinese Mathematics General History, Song–Yuan Volume*]). 第四章. 江苏教育出版社 (di si zhang, Jiangsu jiaoyu chubanshe [Jiangsu education press, chapter 4 184–239]).

Li (Jimin) 李继闵 [1982]. 刘徽对整句股数的研究 (Liu hui dui zheng ju gu shu de yanjiu [Liu Hui's research about Pythagorean triplets]); 科技史文集; 第 8 辑 (keji shi wenji; di 8 ji [*Anthology of the history of Science and Technology* **8** 51–53]).

Li (Jimin) 李继闵 [1990]. 东方数学典籍 — 《九章算术》及其刘徽注研究 (Dongfang shuxue dianji — 'jiu zhang suanshu' ji qi liu hui zhu yanjiu [*A Study on Oriental Mathematical Classic — The Nine Chapters on Mathematical Procedures and Liu Hui's Commentary for It* 1–492]); 陕西人民教育出版社 (Shaanxi renmin jiaoyu chuban she [Shaanxi Peoplpe's Education Press]).

Li (Peiye) 李培业 and Yuan (Min) 袁敏 [2009]. 益古演段释义 (Yigu yanduan shiyi [*Interpretation of the Yigu yanduan*]), 陆西出版集团 (Luxi chuban jituan [Lu xi Publishing Group]), 陕西出版社 (Shaanxi chubanshe [Shaanxi press]).

Li (Yan) 李俨 [1926], 重差术源流及其新注 (Chong cha shu yuanliu ji qi xin zhu [*The origins of the procedure of double difference and a new commentary*]). 学艺第 8 卷第 8 期 (Xueyi di 8 juan di 8 qi) [*Learning Arts* **8**(8) 1–15].

Li (Yan) 李俨 [1955]. 中国史论丛, 第四章 (Zhongguo shi luncong di si zhang [*Discussion on Chinese History*] Chapter 4, 1–365), 科学出版社 (kexue chubanshe [Science Press]).

Li (Yan) 李俨 and Du (Shiran) [1987]. Chinese Mathematics, A Concise History. Translated by Crossley J. N. and Lun W. C. Clarendon Press. Oxford.

Libbrecht (Ulrich) [1973]. Chinese mathematics in the thirteenth century. The Shushu chiu-chang of Chin Chiu-shao. The MIT Press, Cambridge.

Liu (Dun) and Dauben (Joseph) [1993]. The "Qian-Jia School" and its successors. 19th Century Chinese Mathematics. J. W. Dauben and C. J. Scriba: Writing the History of Mathematics, 300–306. Bikhäuser.

Mancosu (Paolo) [2005]. Visualization in logic and mathematics. Mancosu. P, Jorgensen K.F, Pedersen S.A (Eds), Visualization, Explanation and Reasoning Styles in Mathematics. Springer (Netherland).

Martzloff (Jean-Claude) [1987]. Histoire des mathématiques chinoises. Masson. Paris. For English edition: A History of Chinese Mathematics, Springer, 1997.

Mei (Rongzhao) 梅荣照 [1966]. 李冶及其数学著作 (Li Ye ji qi shuxue zhuzuo [*Li Ye and His Mathematical Works*]). 宋元术学史论文集 (Song–Yuan shuxue shi lunwenji [Collection on Song–Yuan History of Mathematics] 104–148), edited by 钱宝琮 Qian Baocong, 科学出版社 (kexue chubanshe [Science Press]).

Mei (Rongzhao) 梅荣照 [1984]. 刘徽的方程理论; 刘徽的句股理论 (Liu hui de fangcheng lilun; liu hui de ju gu lilun [*Liu Hui's theory about Fangcheng, Liu Hui's theory of Right-angled triangles*]); 科学史集刊第 11 集 (kexue shi jikan, di 11 ji [*Anthology of History of Science* (11) 63–76; 77–95.])

Mikami (Yoshio) [1913]. The Development of Mathematics in China and Japan (Ch. 12 Li Yeh 79–84). Leibzig, New Yorkm BG Teubner.

Morgan (Daniel Patrick) [2015]. What good's a text? Textuality, orality, and mathematical astronomy in early imperial China. *Archives internationales d'histoire des sciences* **65**(2) 549–572.

Needham (Joseph) and Wang (Ling) [1954]. Science and Civilisation in China III 42–151. Cambridge.

Needham (Joseph) Colin A. Ronan [1978]. The Shorter Science and Civilisation in China. Cambridge University Press.

Netz (Reviel) [1999]. The Shaping of Deduction in Greek Mathematics. A Study in Cognitive History (Ideas in Context 51). Cambridge University Press.

Pollet (Charlotte) [2012]. Comparison of algebraic practices in medieval China and India. PhD diss. National Taiwan Normal University/Université Paris 7-Diderot. Not published.

Pollet (Charlotte) [2014]. The influence of Qing dynasty editorial work on the modern interpretation of mathematical sources: the case of Li Rui's edition of Li Ye's mathematical treatises. *Science in Context* **27**(3) 385–422. Cambridge University Press.

Pollet (Charlotte) and Ying (Jiaming) [2017]. One quadratic equation, different understandings: the 13th century interpretation by Li Ye and later commentators in the 18th and 19th centuries. *Journal for History of Mathematics. Korean Society for History of Mathematics* **30**(3) 137–162.

Qian (Baocong) 钱宝琮 [1937]. 中国数学中之整数句股形研究 (Zhongguo shuxue zhong zhi zhengshu ju gu xing yanjiu [*A study on the right-angled triangles with integral side lengths in Chinese mathematics*]), 数理杂志第 1 卷第 3 期 (Shuli zazhi. Di 1 juan di 3 qi [Mathematics and Science Magazine **1**(3) 94–112]).

Qian (Baocong) 钱宝琮 [1963]. 算径十书 (Suanjing shishu [*The ten Classical Books in Mathematics*]). 北京, 中华书局 (Beijing, Zhonghua shuju [Chinese Book Compagnie]).

Qian (Baocong) 钱宝琮 [1964]. 中国数学史 (Zhongguo shuxue shi [*History of Chinese mathematics*]). 科学出版社 (Kexue Chubanshe [Science Press]).

Qu (Anjing) [1997]. On hypothenuse diagrams in ancient China. *Centaurus* **39** 193–210.

Robinet (Isabelle) [1979]. Méditation Taoïste. Collection spiritualité. Albin Michel. Paris.

Robinet (Isabelle) [1990]. The place and meaning of the notion of Taiji in Taoist sources prior to the Ming dynasty. *History of Religions* **23**(4) 373–411.

Robinet (Isabelle) [1991]. Histoire du Taoïsme, des origins au XIVe siècle. Patrimoines taoïsme. Cerf.

Robinet (Isabelle) [2012]. Introduction à l'alchimie intérieure taoïste. De l'unité et de la multiplicité, avec une traduction commentée des Versets de l'éveil a la vérité. Patrimoines taoïsme. Cerf. Paris.

Schäfer (Dagmar) [2011]. The Crafting of 10000 Things. Knowledge and Technology in Seventeenth-Cenury China. The University of Chicago Press. Chicago and London.

Sarton (George) [1927]. Introduction to the history of science **II** Part II 627–628.

Te (Gusi) 特古斯 [1990]. 刘益及其佚著《议古根源》(Liu Yi ji qi yi zhu 'Yigu gen yuan' [*Liu Yi and His Lost Work, the 'Yigu genyuan'*]. 李迪 (ed) 数学史研究文集, (shuxueshi yanjiu wenji [*Research on History of Mathematics Series* 56–63]) 九章出版社 (jiu zhang chubanshe [The Nine Chapter Press]).

Tian (Miao) [1999]. Jiegenfang, Tianyuan and Daishu: algebra in Qing China. *Historia Scientarum* **9**(1) 101–119.

Volkov (Alexeï) [1992]. Analogical reasoning in ancient China. Some Examples. Chemla (Eds), Regards obliques sur l'argumentation en Chine. Extreme-Orient, Extreme-Occident **14** 15–48.

Volkov (Alexeï) [1994]. Transformations of geometrical objects in Chinese mathematics and their evolution. Alleton V. and Volkov A. (Eds). Notions et perceptions du changement en Chine: Textes présentés au IXe Congrès de l'Association européenne d'études chinoises (Mémoires de l'Institut des hautes études chinoises XXXVI 133–148.) Collège de France, Institut des hautes études chinoises (1994).

Volkov (Alexeï) [1997]. Science and Taoism: an introduction. *Taiwanese Journal for Philosphy and History of Science* **5**(8) 1–58.

Volkov (Alexeï) [2001]. Le Bacchette. Chemla, Bray, Fu, Huang and Metaile (eds), La scienza in Cina, Storia della Scienza l(8) section I, Vol. II. Enciclopedia Italiana.

Volkov (Alexeï) [2004]. Scientific knowledge in Taoist context : Chen Zhixu's 陈致虚 Commentary on the Scripture of Salvation (Duren jing 度人经). Religion and Chinese Society. Vol II : Taoism and Local Religion in Modern China. Edited by John Lagerwey. A centennial Conference of the Ecole Française d'Extêrme-Orient. The Chinese University of Hong Kong and Ecole Française d'Extêrme-Orient.

Volkov (Alexeï) [2006]. Le raisonnement par analogie dans les mathématiques chinoises du premier millénaire de notre ère. Durand-Richard M-J. L'analogie dans la démarche scientifique. L'Harmattan. Collection Histoire des Sciences, série études.

Volkov (Alexeï) [2007]. Geometrical diagrams in traditional Chinese mathematics. Francesca Bray, Vera Dorofeeva-Lichtmann, Georges Metailie (Eds), Graphics and text in the Production of Technical Knowledge in China. *Sinica Leidensia* **79** 425–460. Brill. Leiden-Boston.

Wang (Ling) [1964]. The date of the Sun Tzu Suan Ching and the Chinese remainder problem, *Actes du Xe Congres International d'histoire des sciences* 489–492 Paris.

Wu (Wenjun) 吴文俊 [1985]. 益古演段. 李冶的数学成就 (Yi gu yan duan. Li ye de shuxue chengjiu. [*The Development of Pieces of Areas according to the Collection Augmenting the Ancient Knowledge, Li Ye's mathematical achievements*]). 中国数学史大系. 第二编. 第三章 (Zhongguo shuxue shi da xi. Di er bian. Di san zhang [*Collection of History of Chinese Mathematics*, Sec. 2, Ch. 3 104–130.])

Xu (Yibao) 徐义保 [1990]. 对《益古集》的复原与研究 (Dui 'yi gu ji' de fuyuan yu yanjiu [*Restoration and Research on the Collection Augmenting the Ancient [knowledge]*]). 数学史研究文集 (Shuxueshi yanjiu wenji [*Research on History of Mathematics Series* 149–165]), edited by Li Di. 九章出版社 (Jiu zhang chubanshe [Nine Chapters Press]).

Xu (Zelin) 徐泽林 and Wei (Xia) 卫霞 [2011]. "演段" 考释—兼论东亚代数演算方式的演变 ('yan duan' kaoshi — jian lun dongya daishu yansuan fangshi de yanbian [*Textual Research on the Expression 'yan duan' — Also Discussing the Evolution of Algebraic Computational Methods in East Asia*]). 自然科学史研究第 30 卷第 2 期 (2011 年): 219–245. (ziran kexueshi yanjiu, [*The History of Natural Sciences* **30**(2) 219–245].

Zhou (Hanguang) 周瀚光 [1987]. (Lun Li Ye de kexue sixiang [*On Li Ye's Scientific Thoughts*]). 吴文俊 Wu Wenjun (eds), 中国数学史论文集 (三) (Zhongguo shuxue shilun wenji (san) [*Colleciton of Papers on History of Chinese Mathematics (Three)* 73–80]). 山东教育出版社 (Shandong jiaoyu chuban she [Shangdong Education Press]).

Zürcher (Erik) [2007]. The Buddhist Conquest of China. The Spread and Adaptation of Buddhism in Early Medieval China, third edition. Leiden, Brill.

Index